ECONOMIC AND SOCIAL COMMISSION
FOR ASIA AND THE PACIFIC
Bangkok, Thailand

# NUCLEAR ENERGY FOR PEACEFUL USES: STATUS AND SALIENT ISSUES FOR REGIONAL CO-OPERATION

ENERGY RESOURCES DEVELOPMENT SERIES
NO. 29

UNITED NATIONS
New York, 1985

ST/ESCAP/374

| UNITED NATIONS PUBLICATION |
| :---: |
| Sales No. : E.86.II.F.4 |
| ISBN 92-1-119405-9 |
| ISSN 0252-4368 |

Price: $US 10.50

# PREFACE

It is with much pleasure and a sense of achievement that volume 29 in the Energy Resources Development Series of ESCAP is offered to the countries in the ESCAP region.

In perusing it, one cannot help but feel that high technology might in certain cases offer quick solutions to some regional development problems by avoiding possible bottle-necks and providing short cuts for the development process. Nuclear technology, while at present perhaps beyond the reach of some of the ESCAP member countries, could in the immediate future conceivably offer such an opportunity, especially through the current global efforts to implement the objectives of the forthcoming United Nations Conference for the Promotion of International Co-operation in the Peaceful Uses of Nuclear Energy.

The present publication provides an overview of the current and potential peaceful uses of nuclear energy which could be harnessed by developing countries to support their development efforts.

I wish to express my sincere thanks and appreciation to all those who have made the publication of this volume possible. My thanks go especially to Mr. Amrik Mehta, Secretary General of the Conference, whose untiring personal efforts will, I am certain, play a significant part enhancing regional and international co-operation in the important field of the peaceful uses of nuclear energy.

S.A.M.S. Kibria
Executive Secretary
ESCAP

# FOREWORD

The importance of the use of nuclear energy for peaceful purposes for the economic and social development of many countries is universally recognized. Equally, the importance of the role of international co-operation in facilitating the introduction and development of peaceful uses of nuclear energy, particularly in the developing countries, cannot be over-emphasized.

In the contemporary world, modern science and technology have deeply and irreversibly altered the pattern of our lives. In stimulating change and innovation and in promoting the birth of new and highly specialized industries, science and technology have brought unprecedented prosperity to the northern part of the globe and, at the same time, raised hopes in the less fortunate and more populous part of the world that it may also aspire to a better standard of human life. Nuclear science and technology can play a significant part in meeting this expectation. Given the increasing energy requirements for economic and social development so much needed in the greater part of the world, and considering that a number of countries have little or no access to other energy sources, the introduction of nuclear power in their energy programmes and other peaceful nuclear applications in food and agriculture, health and medicine, hydrology, industry, etc., could play a vital role in this respect. The question is: how can this concept be transformed into reality on the widest possible scale over a reasonable span of time. There are obvious difficulties and constraints which inhibit the introduction and development of peaceful uses of nuclear energy in many parts of the world. The problems faced are, to a large measure, similar in content. These can be resolved only by fostering international co-operation to the maximum extent possible.

The United Nations Conference for the Promotion of International Co-operation in the Peaceful Uses of Nuclear Energy represents the first international effort of its kind designed exclusively for the purpose of promoting international co-operation in the peaceful uses of nuclear energy for economic and social development. ESCAP has actively participated in the preparations for this important Conference and made a useful contribution through the report of the regional expert group meeting held at Bangkok in January 1985. This issue of the Energy Resources Development Series should also help the countries of the region in focusing attention on the specific purpose, aims and objectives of the Conference and the central issue of the promotion of international co-operation in this field.

AMRIK S. MEHTA
Secretary General
United Nations Conference for the
Promotion of International Co-
operation in the Peaceful Uses
of Nuclear Energy

ESCAP:     The United Nations Economic and Social Commission for Asia
and the Pacific

The term "ESCAP region" is used to include Afghanistan, Australia,
Bangladesh, Bhutan, Brunei, Darussalam, Burma, China, Cook Islands,
Democratic Kampuchea, Fiji, Guam, Hong Kong, India, Indonesia, Iran
(Islamic Republic of), Japan, Kiribati, Lao People's Democratic Republic,
Malaysia, Maldives, Mongolia, Nauru, Nepal, New Zealand, Niue, Pakistan,
Papua New Guinea, the Philippines, Republic of Korea, Samoa, Singapore,
Solomon Islands, Sri Lanka, Thailand, Tonga, Trust Territory of the Pacific
Islands, Tuvalu, Vanuatu and Viet Nam.   The term "developing ESCAP
region" excludes Australia, Japan and New Zealand.

# CONTENTS

# CONTENTS

# SUMMARY

In preparation for the United Nations Conference for the Promotion of International Co-operation in the Peaceful Uses of Nuclear Energy, ESCAP commissioned a survey and organized an expert group meeting in 1984 and 1985. The current volume contains this survey as well as the report and recommendations of the expert group. The report of the expert group meeting is an input to the global preparatory process. It is felt that the information contained in the survey, although perhaps available from other sources, has not been presented before in such concise form to a non-specialist audience. Hence the decision to publish the current volume. At the same time the recommendations of the regional expert group, while part of the global process of preparations for the 1987 United Nations Conference, may also have immediate regional implications and thus deserve the wider dissemination rendered possible by the current volume.

There were ten recommendations in the nuclear power area by the regional experts and six others in the field of general nuclear techniques, radioisotopes and radiation and an overall general recommendation to follow-up implementation at the international level.

There was a specific recommendation concerning the role of ESCAP in promoting nuclear techniques in food and agriculture, animal sciences, industry and health, towards social and economic development as well as a recommendation to monitor progress in international co-operation in the nuclear field via an intergovernmental mechanism to be proposed to the planned 1987 Conference.

# INTRODUCTION

The United Nations Conference for the Promotion of International Co-operation in the Peaceful Uses of Nuclear Energy to be held in March 1987 at Geneva will be the first international effort designed exclusively for the purposes of promoting international co-operation in the peaceful uses of nuclear energy for economic and social development. The deliberations at the Conference will consist of three items: one on principles universally acceptable for international co-operation in the peaceful uses of nuclear energy and appropriate ways and means for promotion of such co-operation as envisaged in General Assembly resolution 32/50 of 8 December 1977 and in accordance with mutually acceptable considerations of non-proliferation; one on the role of nuclear power for social and economic development; one on the role of other peaceful applications of nuclear energy such as in food and agriculture, health and medicine, hydrology, industry, etc. for social and economic development. The General Assembly has requested all States, the International Atomic Energy Agency (IAEA), the specialized agencies and other relevant organizations of the United Nations system to co-operate and to make contributions to the preparations for the Conference.

For the Preparatory Committee of the Conference, regional commissions including ESCAP were requested to provide reports containing practical recommendations for regional measures to be considered by the 1987 Conference.

In its regional involvement, guided by the General Assembly's resolutions relating to the Conference, ESCAP conducted a preliminary study on relevant nuclear energy topics in the region and later ESCAP furnished to member States in the region, a provisional paper NR/ICPUNE/1 to serve as a background document for participating representatives prior to the convening of an expert group meeting in the region.[1] ESCAP also sent detailed guide-lines to member States so that the expert group meeting would be able to evaluate the current situation and projected development in all aspects in the region.

---

[1] The Secretariat was assisted by Dr. Svasti Srisukh, former Secretary-General of the Thai Government's Office of Atomic Energy for Peace, in the preparation of this document and the subsequent technical editing of the current volume, as well as Drs. A.K. Kaul and P.P.G. Lionel Siriwardene as consultants for the agricultural and industrial applications of nuclear techniques at the regional expert group meeting.

The document covered broadly the two agenda items of the Conference namely, nuclear power and applications of nuclear techniques in aspects of agriculture and food preservation, health and medicine, hydrology and industry. However, certain topics such as the nuclear fuel cycle, nuclear safety, radioactive waste disposal and the environmental impact of nuclear power might be too specialized or lengthy to be included. In the document it was found most appropriate to incorporate the development of the co-operation in nuclear science and technology interregionally and internationally, in particular the Regional Co-operative Agreement for Research, Development and Training related to Nuclear Science and Technology since the latter is of special importance for this region. The document which comprises the first three parts of this publication served the purpose of providing experts of varied disciplines to the expert group meeting with general background information on development. Also it would furnish information to national non-specialist administrators of some developing countries in the ESCAP region which have not yet adopted any nuclear techniques applications to take note of development in different fields for future possible regional co-operation.

The Regional Expert Group Meeting was held at Bangkok from 14 to 17 January 1985 with the participation of Mr. Amrik S. Mehta, Personal Representative of the Secretary-General of the United Nations and Secretary-General of the Conference. The experts undertook the review of the present status and future prospects in the light of specific regional needs and priorities, the assessment of main problems and constraints experienced or likely to be faced, and the suggestion of specific initiatives aimed at overcoming such constraints concerning introduction and development of nuclear power in the region and in connection with other peaceful applications of nuclear science and technology in the region. Practical measures and effective ways and means of promoting co-operation in the field, at regional and international levels were taken into consideration. The report of the regional expert group meeting for the Conference together with recommendations was adopted on 17 January 1985 (NR/ICPUNE/2) and forms part four of the current volume.

An annex to this volume contains references, data sources and abbreviations.

Part One

**NUCLEAR POWER AND NUCLEAR RAW MATERIALS**

# 1. NUCLEAR POWER

If economic development is to take place in the developing countries of the ESCAP region, rapid growth in energy supplies and electric power production to meet the increasing energy requirements is a must. During the past decade of energy crisis, this continuing need has been evidenced by sustained growth in electric power, both production and development, in developing countries in the ESCAP region, in spite of obstacles. This trend of steady growth will no doubt continue for quite some time. In an ESCAP secretariat note[1] an attempt was undertaken to make long-term projections (up to 1995) of energy supply developments as a technical extrapolation of past development trends. However, it is to be borne in mind that the countries in the ESCAP region cover a wide range of developmental situations: some are highly industrialized, some are partly industrialized and some are based on agriculture. This situation governs the pattern of electric power demand and growth, and the search for a regional approach is unavoidably limited in scope and plagued by over-aggregation. Nevertheless, most of the countries are developing countries and are facing common energy problems. Energy is the *sine qua non* for the achievement of their socio-economic objectives.

Most of the electric power industries in the region are still highly dependent on imported petroleum-based fuels, except for a few countries which have their own energy reserves. Table 1.1 shows the generation of electricity in public utilities and self-generating industries in the ESCAP region by type of plant, 1971-1982. It can be observed that electricity generated from thermal energy has dominated the electric power industries in this region during the past decade (1971-1980) or even before, accounting for over 70 per cent of total generation. The total electricity production for 1981 and 1982 for the ESCAP region was 1,299,537 and 1,343,456 gigawatt-hours (GWh) as compiled from available reports. The shares for nuclear electricity production increased to 8.3 per cent for 1982 though figures for Pakistan and Taiwan (province of the People's Republic of China) were not included.

Among the primary energy sources for thermal electricity generation, petroleum has by far been the preferred fuel in most countries because of the relatively shorter time required to bring new generating capacities onstream, ease of handling, availability of technical skills, and the relative cheapness of petroleum over the past years (pre-1973). The changed relative price of petroleum has spurred interest in

the use of alternative sources of primary energy, including forms of renewable energy. In the current world-wide energy transition, the scope for rapid substitution of fuel for generation of electricity may be important. Increased use of coal and natural gas is taking place in thermal generation. At present India, Pakistan and the Republic of Korea, among the developing countries in the region, have nuclear power reactors.

Tables 1.2 and 1.3 show the statistics on nuclear power in the ESCAP region. Although development in the region as a whole is significant, it is somewhat slow except in Japan (a developed country). The Republic of Korea, a new member in the nuclear power field, has an ambitious new development programme, and the programme of India has been under way for a number of years. Table 1.4 shows nuclear power plants in developing countries in Asia and the Pacific and in Japan.

Over the medium term, nuclear power offers a substitution for the use of oil for generation of electricity and it represents for many countries, deficient not only in oil and gas but also in coal, a possible viable alternative.

Further developments in nuclear power generation in the region are limited by a number of constraints. These might include (a) electricity grid sizes and unavailability of small power reactors; (b) lack of manpower and infrastructure; (c) lack of capital resources; (d) uncertainties in supply (fuel, equipment, spare parts); (e) spent fuel reprocessing and waste disposal problems; and (f) environmental and socio-political acceptance.

However, nuclear power technology would be new to some countries and they would have to depend on imported technology. Co-operation is therefore necessary in technology transfer in order to raise the capability of these countries at least to the technology adaptation stage, if not the stage of manufacturing of components. Access to information and the results of research activities should be ensured. In the region Japan has developed industrial self-sufficiency and could appear as an exporter of nuclear power plants although it has not done so. India is almost self-sufficient in nuclear power plant production.

It is a common belief that reactors smaller than 600 megawatts (MW) are not economically viable. Under this limitation, there are only a few utility systems in the ESCAP region which could accommodate plants of over 600 MW. As a result, some interested countries with relatively low system capacity are losing interest. This is

---

[1]  "Current and projected energy situation in the ESCAP region" (E/ESCAP/NR. 8/18).

### Table 1.1. Generation of electricity in public electric utilities and self-generating industries in the ESCAP region by type of plant, 1971 to 1982[a]
(In millions of kilowatt hour (KWh))

| Year | Total | Hydro | Thermal | | | | Nuclear |
|------|-------|-------|---------|-------|---------------------|-------------|---------|
| | | | Total | Steam | Internal combustion | Gas turbine | |
| 1971 | 579 951[b] | 157 676 | 412 477 | 397 116 | 13 433 | 1 928 | 9 199 |
| (percentage) | (100) | (27.2) | (71.1) | (68.5) | (2.3) | (0.3) | (1.6) |
| 1972 | 625 392[b] | 159 861 | 454 320 | 436 905 | 15 048 | 2 367 | 10 611 |
| (percentage) | (100) | (25.6) | (72.6) | (69.9) | (2.4) | (0.4) | (1.7) |
| 1973 | 702 015 | 144 667 | 544 842 | 534 559 | 7 721 | 2 562 | 12 506 |
| (percentage) | (100) | (20.6) | (77.6) | (76.1) | (1.0) | (0.4) | (1.8) |
| 1974 | 705 399 | 161 449 | 521 386 | 505 700 | 12 620 | 3 066 | 22 565 |
| (percentage) | (100) | (22.9) | (73.9) | (71.7) | (1.8) | (0.4) | (3.2) |
| 1975 | 742 139 | 172 229 | 541 500 | 526 019 | 12 344 | 3 137 | 28 411 |
| (percentage) | (100) | (23.2) | (73.0) | (70.9) | (1.7) | (0.4) | (3.8) |
| 1976 | 790 753 | 173 642 | 579 745 | 562 650 | 10 044 | 7 051 | 37 365 |
| (percentage) | (100) | (22.0) | (73.3) | (71.2) | (1.4) | (0.9) | (4.7) |
| 1977 | 831 929 | 158 148 | 639 779 | 621 662 | 10 884 | 7 233 | 34 002 |
| (percentage) | (100) | (19.0) | (76.9) | (74.7) | (1.3) | (0.9) | (4.1) |
| 1978 | 874 962 | 161 081 | 652 786 | 629 691 | 13 988 | 9 107 | 61 094 |
| (percentage) | (100) | (18.4) | (74.6) | (72.0) | (1.6) | (1.0) | (7.0) |
| 1979 | 1 195 834[c] | 221 367 | 898 128 | 873 897 | 13 663 | 10 568 | 76 422 |
| (percentage) | (100) | (18.5) | (75.1) | (73.1) | (1.1) | (0.9) | (6.4) |
| 1980 | 1 218 676[d] | 233 027 | 894 447 | 870 389 | 11 994 | 12 064 | 88 945 |
| (percentage) | (100) | (19.1) | (71.4) | (71.4) | (1.0) | (1.0) | (7.3) |
| 1981 | 1 299 537 | 272 889 | 932 910 | 902 895 | 14 670 | 11 496 | 93 738[e] |
| (percentage) | (100) | (21.0) | (73.4) | (69.5) | (1.1) | (0.9) | (7.2) |
| 1982 | 1 343 456 | 275 713 | 956 873 | 923 303 | 15 851 | 12 937 | 110 870[e] |
| (percentage) | (100) | (20.5) | (71.2) | (68.7) | (1.2) | (1.0) | (8.3) |

*Source:* *Electric Power in Asia and the Pacific, 1971-1972, 1973-1974, 1975-1976, 1977-1978, 1979-1980* and *1981-1982,* United Nations ESCAP publications.

[a] Data cover most countries for which information was available. Data for 1971 to 1978 exclude China, but data for China 1982 include steam only.

[b] Including unclassified generation of 600 million kWh in Pakistan.

[c] Including unclassified generation of 82 million kWh in Sri Lanka.

[d] Including unclassified generation of 2,257 million kWh in the northern part of Viet Nam.

[e] Excluding generation in Pakistan and Taiwan (province of the People's Republic of China).

*Note:* Geothermal was about 0.3 per cent for 1981 and 1982.

**Table 1.2. Installed nuclear power generating capacities of public electric utilities
and self-generating industries**
(In megawatts (MW))

| Country or area and year | | Public electric utilities | | | Self-generating industries | Total |
|---|---|---|---|---|---|---|
| | | Publicly owned | Privately owned | Total | | |
| ESCAP region[a] | 1979 | 1 352 | 14 952 | 16 304 | 178 | 16 482 |
| | 1980 | 1 352 | 15 511 | 16 863 | 177 | 17 040 |
| | 1981 | 1 572 | 16 077 | 17 649 | 178 | 17 827 |
| | 1982 | 2 251 | 17 177 | 19 428 | 165 | 19 593 |
| India | 1979 | 640 | – | 640 | – | 640 |
| | 1980[b] | 640 | – | 640 | – | 640 |
| | 1981 | 860 | – | 860 | – | 860 |
| | 1982 | 860 | – | 860 | – | 860 |
| Japan | 1979 | – | 14 952 | 14 952 | 178 | 15 130 |
| | 1980 | – | 15 511 | 15 511 | 177 | 15 688 |
| | 1981 | – | 16 077 | 16 077 | 178 | 16 255 |
| | 1982 | – | 17 177 | 17 177 | 165 | 17 342 |
| Republic of Korea | 1979 | 587 | – | 587 | – | 587 |
| | 1980 | 587 | – | 587 | – | 587 |
| | 1981 | 587 | – | 587 | – | 587 |
| | 1982 | 1 266 | – | 1 266 | – | 1 266 |
| Pakistan | 1979 | 125 | – | 125 | – | 125 |
| | 1980 | 125 | – | 125 | – | 125 |
| | 1981 | 125 | – | 125 | – | 125 |
| | 1982 | 125 | – | 125 | – | 125 |

*Source:* Electric Power in Asia and the Pacific, 1979 and 1980, and 1981 & 1982 (ESCAP, Bangkok).

[a] For nuclear power plants in operation in India, Japan, Pakistan and the Republic of Korea.

[b] Preliminary figures not confirmed or latest figures not available.

### Table 1.3. Nuclear electric power generation in public electric utilities
### and self-generating industries
(In millions of kilowatt hours (kWh))

| Country or area and year | | Public electric utilities | | | Self-generating industries | Total |
|---|---|---|---|---|---|---|
| | | Publicly owned | Privately owned | Total | | |
| ESCAP region[a] | 1979 | 6 029 | 69 344 | 75 373 | 1 049 | 76 422 |
| | 1980 | 6 354 | 82 009 | 88 363 | 582 | 88 945 |
| | 1981 | 5 918 | 87 231 | 93 149 | 589 | 93 738 |
| | 1982 | 8 440 | 101 835 | 110 275 | 595 | 110 870 |
| India | 1979 | 2 877 | – | 2 877 | – | 2 877 |
| | 1980[b] | 2 877 | – | 2 877 | – | 2 877 |
| | 1981[b] | 3 021 | – | 3 021 | – | 3 021 |
| | 1982[b] | 4 663 | – | 4 663 | – | 4 663 |
| Japan | 1979 | – | 69 344 | 69 344 | 1 049 | 70 393 |
| | 1980 | – | 82 009 | 82 009 | 582 | 82 591 |
| | 1981 | – | 87 231 | 87 231 | 589 | 87 820 |
| | 1982 | – | 101 835 | 101 835 | 595 | 102 430 |
| Republic of Korea | 1979 | 3 152 | – | 3 152 | – | 3 152 |
| | 1980 | 3 477 | – | 3 477 | – | 3 477 |
| | 1981 | 2 897 | – | 2 897 | – | 2 897 |
| | 1982 | 3 777 | – | 3 777 | – | 3 777 |

*Source:*   Electric Power in Asia and the Pacific, 1979 and 1980, and 1981 and 1982 (ESCAP, Bangkok)

[a]   Including nuclear power stations of India, Japan, Pakistan and the Republic of Korea. No current data were available for Pakistan. Regional total excludes data for Pakistan.

[b]   Preliminary figures not confirmed or latest figures not available.

### Table 1.4. Nuclear power plants in the ESCAP region

| Code IAEA | Country or area Name | Type | Capacity (Megawatts (MWe)) Net | Gross | Operator | Contractor | Construction start | First criticality | Commercial operation |
|---|---|---|---|---|---|---|---|---|---|
| **CHINA** | | | | | | | | | |
| CN-1 | Qinshan | PWR | 300 | | | | 1983-06 | 1988 | |
| CN-2 | Guandong-1 | PWR | 900 | | | | 1984 | | |
| CN-3 | Guandong-2 | PWR | 900 | | | | 1984 | | |
| CN-4 | Jinshan-1 | PWR | | 450 (Heat only) | | | | | |
| CN-5 | Jinshan-2 | PWR | | 450 (Heat only) | | | | | |
| **TAIWAN (province of the People's Republic of China)** | | | | | | | | | |
| TW-1 | Chin-Shan-1 | BWR | 604 | 636 | TPC | GE. | 1972-06 | 1977-10 | 1978-12 |
| TW-2 | Chin-Shan-2 | BWR | 604 | 636 | TPC | GE. | 1973-12 | 1978-11 | 1979-09 |
| TW-3 | Kuosheng-1 | BWR | 951 | 985 | TPC | GE. | 1975-11 | 1981-02 | 1981-12 |
| TW-4 | Kuosheng-2 | BWR | 951 | 985 | TPC | GE. | 1976-03 | 1982-03 | 1983-03 |
| TW-5 | Maanshan-1 | PWR | 907 | 951 | TPC | WEST. | 1978-08 | | 1984 |

## Table 1.4. *(continued)*

| Code IAEA | Country or area Name | Type | Capacity (Megawatts (MWe)) Net | Gross | Operator | Contractor | Construction start | First criticality | Commercial operation |
|---|---|---|---|---|---|---|---|---|---|
| TW-6 | Maanshan-2 | PWR | 907 | 951 | TPC | WEST. | 1979-02 | | 1985 |
| TW-7 | Taipower-7 | PWR | 910 | 951 | TPC | . | | | |
| TW-8 | Taipower-8 | PWR | 910 | 951 | TPC | | | | |
| TW-9 | Taipower-9 | | 1150 | 1200 | TPC | | | | |
| TW-10 | Taipower-10 | | 1150 | 1200 | TPC | | | | |
| | **INDIA** | | | | | | | | |
| IN-1 | Tarapur-1 | BWR | 198 | 210 | DAE | GE. | 1964-10 | 1969-02 | 1969-10 |
| IN-2 | Tarapur-2 | BWR | 198 | 210 | DAE | GE. | 1964-10 | 1969-02 | 1969-10 |
| IN-3 | Rajasthan-1 | PHWR | 207 | 220 | DAE | AECL. | 1964-12 | 1972-08 | 1973-12 |
| IN-4 | Rajasthan-2 | PHWR | 207 | 220 | DAE | AECL. | 1968-06 | 1980-10 | 1981-04 |
| IN-5 | Kalpakkam-1 | PHWR | 220 | 235 | DAE | DAE | 1970-02 | 1983-07 | 1983-12 |
| IN-6 | Kalpakkam-2 | PHWR | 220 | 235 | DAE | DAE | 1971-05 | 1984-01 | 1984-12 |
| IN-7 | Narora-1 | PHWR | 220 | 235 | DAE | DAE | 1976-01 | 1986-07 | 1987-06 |
| IN-8 | Narora-2 | PHWR | 220 | 235 | DAE | DAE | 1976-01 | 1987-06 | 1988-06 |
| IN-9 | Kakarpur-1 | PHWR | 220 | 235 | DAE | DAE | 1983 | 1990 | 1991 |
| IN-10 | Kakarpur-2 | PHWR | 220 | 235 | DAE | DAE | 1983 | 1991 | 1992 |
| IN-11 | Kakarpur-3 | PHWR | 220 | 235 | DAE | DAE | | | |
| IN-12 | Kakarpur-4 | PHWR | 220 | 235 | DAE | DAE | | | |
| IN-13 | | PHWR | 220 | 235 | DAE | DAE | | | |
| IN-14 | | PHWR | 220 | 235 | DAE | DAE | | | |
| | **REPUBLIC OF KOREA** | | | | | | | | |
| KR-1 | Ko-Ri-1 | PWR | 556 | 587 | KEPCO | WEST. | 1970-09 | 1977-06 | 1978-04 |
| KR-2 | Ko-Ri-2 | PWR | 605 | 650 | KEPCO | WEST. | 1977-05 | 1983-04 | 1983-07 |
| KR-3 | Wolsung-1 | PHWR | 628 | 678 | KEPCO | AECL. | 1977-05 | 1982-11 | 1983-04 |
| KR-4 | Ko-Ri-3 | PWR | 900 | 950 | KEPCO | WEST. | 1979-06 | 1984-05 | 1984-09 |
| KR-5 | Ko-Ri-4 | PWR | 900 | 950 | KEPCO | WEST. | 1979-06 | 1985-05 | 1985-09 |
| KR-6 | | | | | | | | | |
| KR-7 | Yeonggwang-1 | PWR | 912 | 950 | KEPCO | WEST. | 1980-03 | 1985-11 | 1986-03 |
| KR-8 | Yeonggwang-2 | PWR | 912 | 950 | KEPCO | WEST. | 1980-03 | 1985-11 | 1986-03 |
| KR-9 | Uljin-1 | PWR | 925 | 950 | KEPCO | WEST. | 1982-10 | 1988 | |
| KR-10 | Uljin-2 | PWR | 925 | 950 | KEPCO | WEST. | 1982-10 | 1989 | |
| KR-11 | | | 864 | 900 | KEPCO | | | | |
| KR-12 | | | 864 | 900 | KEPCO | | | | |
| KR-13 | | | 864 | 900 | KEPCO | | | | |
| KR-14 | | | 864 | 900 | KEPCO | | | | |
| | **PAKISTAN** | | | | | | | | |
| PK-1 | Kanupp | PHWR | 125 | 137 | PAEC | CGE. | 1966-08 | 1971-08 | 1972-10 |
| PK-2 | Chasnupp | PWR | 900 | | PAEC | | | | |
| | **PHILIPPINES** | | | | | | | | |
| PH-1 | PNPP-1 | PWR | 621 | 652 | RPNPC | WEST. | 1976-03 | 1984-10 | 1985-01 |
| | **JAPAN** | | | | | | | | |
| JP-2 | Tokai-1 | GCR | 159 | 166 | JAPCO | GEC. | 1960-07 | 1965-05 | 1966-07 |
| JP-3 | Tsuruga-1 | BWR | 341 | 357 | JAPCO | GE. | 1966-04 | 1969-10 | 1970-03 |
| JP-4 | Mihama-1 | PWR | 320 | 340 | Kepco | WEST. | 1967-08 | 1970-07 | 1970-11 |
| JP-5 | Fukushima-I-1 | BWR | 439 | 460 | TEPCO | GE. | 1966-12 | 1970-07 | 1971-03 |
| JP-6 | Mihama-2 | PWR | 470 | 500 | Kepco | M | 1968-12 | 1972-04 | 1972-07 |
| JP-7 | Shimane-1 | BWR | 439 | 460 | CECO | Hitachi | 1970-02 | 1973-06 | 1974-03 |
| JP-8 | Takahama-1 | PWR | 780 | 826 | Kepco | WEST. | 1970-04 | 1974-03 | 1974-11 |
| JP-9 | Fukushima-I-2 | BWR | 760 | 784 | TEPCO | GE. | 1969-05 | 1973-11 | 1974-07 |
| JP-10 | Fukushima-I-3 | BWR | 760 | 784 | TEPCO | Toshiba | 1970-10 | 1974-08 | 1976-03 |
| JP-11 | Hamaoka-1 | BWR | 515 | 540 | Chubu | Toshiba | 1971-03 | 1974-06 | 1976-03 |

**Table 1.4.** *(continued)*

| Code IAEA | Country or area Name | Type | Capacity (Megawatts (MWe)) Net | Gross | Operator | Contractor | Construction start | First criticality | Commercial operation |
|---|---|---|---|---|---|---|---|---|---|
| JP-12 | Genkai-1 | PWR | 529 | 559 | Kyushu | M | 1971-03 | 1975-01 | 1975-10 |
| JP-13 | Takahama-2 | PWR | 780 | 826 | Kepco | M | 1971-02 | 1974-12 | 1975-11 |
| JP-14 | Mihama-3 | PWR | 780 | 826 | Kepco | M | 1972-07 | 1976-01 | 1976-12 |
| JP-15 | Ohi-1 | PWR | 1120 | 1175 | Kepco | WEST. | 1972-10 | 1977-12 | 1979-03 |
| JP-16 | Fukushima-I-4 | BWR | 760 | 784 | TEPCO | Hitachi | 1972-09 | 1978-01 | 1978-10 |
| JP-17 | Fukushima-I-5 | BWR | 760 | 784 | TEPCO | Toshiba | 1971-12 | 1977-08 | 1978-04 |
| JP-18 | Fukushima-I-6 | BWR | 1067 | 1100 | TEPCO | GE. | 1973-05 | 1979-01 | 1979-10 |
| JP-19 | Ohi-2 | PWR | 1120 | 1175 | Kepco | WEST. | 1972-11 | 1978-09 | 1979-12 |
| JP-20 | Fugen ATR | HWLWR | 148 | 165 | PNC | Hitachi | 1970-12 | 1978-03 | 1979-03 |
| JP-21 | Tokai-2 | BWR | 1080 | 1100 | JAPCO | GEC. | 1973-03 | 1978-01 | 1978-11 |
| JP-22 | Onagawa-1 | BWR | 497 | 524 | Tohoku | Toshiba | 1979-12 | 1983-10 | 1984-06 |
| JP-23 | Ikata-1 | PWR | 538 | 566 | Shikoku | M | 1973-06 | 1977-01 | 1977-09 |
| JP-24 | Hamaoka-2 | BWR | 814 | 840 | Chubu | Toshiba | 1974-03 | 1978-03 | 1978-11 |
| JP-25 | Fukushima-II-1 | BWR | 1067 | 1100 | TEPCO | Toshiba | 1975-11 | 1981-06 | 1982-04 |
| JP-26 | Fukushima-II-2 | BWR | 1067 | 1100 | TEPCO | Hitachi | 1979-02 | 1983-04 | 1984-01 |
| JP-27 | Genkai-2 | PWR | 529 | 559 | Kyushu | M | 1976-06 | 1980-05 | 1981-03 |
| JP-28 | Sendai-1 | PWR | 846 | 890 | Kyushu | M | 1979-01 | 1983-08 | 1984-07 |
| JP-29 | Takahama-3 | PWR | 830 | 870 | Kepco | M | 1980-12 | 1984-06 | 1985-02 |
| JP-30 | Takahama-4 | PWR | 830 | 870 | Kepco | M | 1980-12 | 1984-12 | 1985-08 |
| JP-31 | Monju | FBR | 250 | 280 | PNC | MHI | 1985-06 | 1991 | |
| JP-32 | Ikata-2 | PWR | 538 | 566 | Shikoku | M | 1987-02 | 1981-07 | 1982-03 |
| JP-33 | Kashiwazaki-1 | BWR | 1067 | 1100 | TEPCO | Toshiba | 1978-12 | 1984-12 | 1985-10 |
| JP-34 | Tsuruga-2 | PWR | 1115 | 1160 | JAPCO | M | 1982-04 | 1986-10 | 1987-06 |
| JP-35 | Fukushima-II-3 | BWR | 1067 | 1100 | TEPCO | Toshiba | 1980-12 | 1984-11 | 1985-07 |
| JP-36 | Hamaoka-3 | BWR | 1066 | 1100 | Chubu | Toshiba | 1982-11 | 1986-09 | 1987 |
| JP-37 | Sendai-2 | PWR | 846 | 890 | Kyushu | M | 1982-05 | 1985-04 | 1986-03 |
| JP-38 | Fukushima-II-4 | BWR | 1067 | 1100 | TEPCO | Hitachi | 1980-12 | 1985-05 | 1986-02 |
| JP-39 | Kashiwazaki-2 | BWR | 1067 | 1100 | TEPCO | Toshiba | 1983-10 | | 1990-10 |
| JP-40 | Kashiwazaki-5 | BWR | 1067 | 1100 | TEPCO | Hitachi | 1983-10 | | 1990/91 |
| JP-41 | Shimane-2 | BWR | 791 | 820 | CECO | Hitachi | 1984 | | 1988-09 |
| JP-42 | Maki-1 | BWR | 796 | 825 | Tohoku | | 1986-12 | | 1991-12 |
| JP-43 | Tomari-1 | PWR | 550 | 579 | HEPCO | | 1984-03 | 1988-04 | 1989-03 |
| JP-44 | Tomari-2 | PWR | 550 | 579 | HEPCO | | 1984-03 | 1989-09 | 1990-08 |
| JP-45 | Genkai-3 | PWR | 1127 | 1180 | Kyushu | | 1985-02 | | |
| JP-46 | Genkai-4 | PWR | 1127 | 1180 | Kyushu | | 1985-02 | | |
| JP-47 | Ikata-3 | PWR | 846 | 890 | Shikoku | | 1985-08 | | |
| JP-48 | Onagawa-2 | BWR | 796 | 825 | Tohoku | | 1987-02 | | |
| JP-49 | Onagawa-3 | BWR | 796 | 825 | Tohoku | | | | |

Abbreviations for reactor types:

 BWR  Boiling light-water-cooled and moderated reactor
 FBR   Fast breeder reactor
 GCR   Gas-cooled, graphite-moderated reactor
 HWLWR Heavy-water-moderated, boiling light-water-cooled reactor
 PHWR  Pressurized heavy-water-moderated and cooled reactor
 PWR   Pressurized light-water-moderated and cooled reactor

*Source:* Nuclear power reactors in the world, Reference data series No. 2, (International Atomic Energy Agency, April 1984).

one of the constraints in the promotion of nuclear power development among the developing countries. The installed generating capacity of various countries in the ESCAP region is shown in table 1.5. In this field, specialized agencies and manufacturers may devote efforts to supplying smaller nuclear power plants for the benefit of the developing countries. Although such plants are commercially offered, the prices quoted have not lead to sufficient orders of such plants to bring down unit costs.

Because of the availability of data for China for the first time (from 1979), the regional figures estimated and/or projected earlier have changed. However, installed generating capacity in developing countries (including China) in 1980 was 140,962 MW. Thus, based on the World Bank's estimate of 7.3 per cent of the total installed capacity by 1990 for nuclear power,[2] the total installed nuclear capacity may be in the order of 10 gigawatts (GW) by that time (International Atomic Energy Agency (IAEA) estimate : 14-15 GW).

Taking into account the rapid expansion programme in the Republic of Korea, the current construction programme in India and the Philippines' launching of a nuclear programme, this figure is likely to be slightly exceeded. If timely start-up does take place for the projects under consideration by those and other countries, total nuclear generating capacity may rise as high as 42 to 45 GW (IAEA estimate: 41-47 GW) by the turn of the century.

To achieve this, huge capital investments would be required. Based on the estimated per kilowatt (KW) investment cost of $US 1,600 (for large multiple units) to $US 2,200 (for single small units),[3] on an average of about $US 80 billion to 85 billion would be required by the developing countries in the ESCAP region. If developed, nuclear power would represent only about 14 to 15 per cent (IAEA estimate: 8 per cent in 1990) of the total projected installed electricity generating capacity. If immediate steps are taken to obtain capital flows for the development of nuclear electricity, the expected nuclear capacity may help to some extent to lower the demand for petroleum. The projection takes into consideration the rapid development of hydropotential; therefore, any unmet capacity will have to be met by thermal power. Although increased quantities of coal and gas for power generation have also been assumed, their alternative uses (e.g. as a substitute for liquid fuels and for petro-chemical products) may limit power generation availability.

Apart from the techno-economic problems, most of the developing countries have a general shortage of skilled manpower. Therefore, in planning any development in this specialized technology, manpower development is a prerequisite and should be part of any bilateral or multilateral co-operation in this field. International specialized organizations (e.g. IAEA) could also make special assistance programmes available.

An extremely sensitive issue is spent fuel reprocessing and waste disposal. Although the quantity of spent fuel poses no danger at the moment because of the small scale of these activities, there are signs that in the full-scale development of nuclear reactors this may become a problem. With the limited scope for reprocessing spent fuel in countries in the region, it must be either kept in proper storage as an interim arrangement or sent abroad for reprocessing. Transportation of spent fuel and waste may prove in the future to be a major obstacle to foreign reprocessing options. In any case, the problem of permanent disposal of high-level wastes remains. Those Asian countries with the largest nuclear programmes seem to have a fairly limited potential for permanent waste disposal of high-level wastes. Thus, a long-term programme for waste management is very important for the nuclear countries in the region.

The sea bed and other international disposal options for high-level waste on a co-operative (global or regional) basis will have to be studied and explored before the spent fuel interim storage or other storage pools become crowded.

Although most of the nuclear reactors in operation or under construction in the ESCAP region are light water reactors, some countries of the region, such as India and Japan, are trying to explore systems like breeders or a combination of breeders and converters that utilize the extensive thorium resources, and to reduce the problems of waste disposal. Further research along these lines is expected.

At the moment socio-political acceptance of nuclear power, even for peaceful uses such as electricity generation, is closely related to public safety, pollution and other environmental constraints although risks associated with nuclear power operation are assessed as low compared with risks associated with other sources, even large-scale hydroelectric power.[4] Public opinion always focuses on the dangers of nuclear industries. The low level of risk associated with nuclear energy should be publicized and people should be informed about safety features so as to increase acceptance. Nevertheless, safety features must always be enhanced in reactor design so as to reduce the consequences of any accident to acceptable levels, assuming that such accidents may occur in spite of all the measures taken to prevent them. Internationally agreed criteria should be imposed on all installations prior to granting any license for construction and operation of reactors.

---

[2] World Bank, *Energy in the Developing Countries* (1980).

[3] *Ibid.*

[4] Louis Neel, "The Realism of Nuclear Power", *World Science News,* vol. XIX, No. 14 (30 March-6 April 1982), p. 12.

**Table 1.5. Installed electricity generating capacities in the ESCAP region, 1981 and 1982.**
(In megawatts (MW))

| | | Installed generating capacity (MW) | | | | | | |
|---|---|---|---|---|---|---|---|---|
| | | Hydroelectric | Nuclear | Geothermal | Steam | Diesel | Gas turbines | Total | Increase over previous year (percentage) |
| Afghanistan | 1981 | 261.5 | .. | .. | .. | 56.8 | 55.6 | 373.9 | .. |
| | 1982 | 265.3 | .. | .. | 52.3 | 58.0 | 58.0 | 433.6 | .. |
| Australia | 1981 | 6 253.2 | .. | .. | 18 396.8 | 325.4 | 963.0 | 25 938.4 | 7.1 |
| | 1982 | 6 332.8 | .. | .. | 20 083.8 | 341.1 | 1 258.0 | 28 015.7 | 8.3 |
| Bangladesh | 1981 | 80.0 | .. | .. | 398.8 | 282.5 | 247.4 | 1 008.7 | .. |
| | 1982 | 130.0 | .. | .. | 398.8 | 309.7 | 242.0 | 1 080.5 | 7.1 |
| Bhutan | 1981 | 3.5 | .. | .. | .. | 14.2 | .. | 17.7 | .. |
| | 1982 | 3.5 | .. | .. | .. | 14.3 | .. | 17.8 | 0.5 |
| Burma[a] | 1981 | 169.0 | .. | .. | 243.0 | 69.0 | 214.0 | 695.0 | 6.9 |
| China | 1981 | 18 285.0 | .. | .. | 44 945.0 | .. | .. | 63 230.0 | 1.9 |
| Cook Islands | 1981 | .. | .. | .. | .. | 6.0 | .. | 6.0 | .. |
| | 1982 | .. | .. | .. | .. | 6.3 | .. | 6.3 | 5.0 |
| Fiji | 1981 | .. | .. | .. | 20.0 | 96.4 | .. | 116.4 | (0.3) |
| | 1982 | .. | .. | .. | 20.0 | 97.2 | .. | 117.2 | 0.7 |
| Hong Kong | 1981 | .. | .. | .. | 3 085.0 | 6.3 | 631.0 | 3 722.3 | 6.9 |
| | 1982 | .. | .. | .. | 3 810.0 | 6.3 | 631.0 | 4 447.3 | 19.5 |
| India[a] | 1981 | 12 174.0 | 860.0 | .. | 22 350.0 | .. | .. | 35 384.0 | 6.8 |
| | 1982 | 13 058.0 | 860.0 | .. | 24 890.0 | .. | .. | 38 808.0 | 9.7 |
| Indonesia[a] | 1981 | 1 313.0 | .. | .. | 1 476.0 | 2 233.0 | 797.0 | 5 819.0 | 7.8 |
| | 1982 | 1 355.0 | .. | 30.0 | 1 676.0 | 2 483.0 | 817.0 | 6 361.0 | 7.2 |
| Islamic Republic | 1981 | 1 804.0 | .. | .. | 4 423.0 | 2 430.0 | 3 175.0 | 11 832.0 | 11.6 |
| of Iran | 1982 | 1 804.0 | .. | .. | 4 423.0 | 2 480.0 | 3 201.0 | 11 908.0 | 0.6 |
| Japan | 1981 | 31 599.0 | 16 255.0 | 131.0 | 98 659.0 | 1 396.0 | 2 001.0 | 150 041.0 | 4.4 |
| | 1982 | 33 313.0 | 17 342.0 | 180.0 | 100 347.0 | 1 434.0 | 2 029.0 | 154 645.0 | 3.1 |
| Malaysia: | | | | | | | | | |
| Peninsular | 1981 | 632.8 | .. | .. | 1 330.0 | 521.0 | 100.0 | 2 583.8 | 1.8 |
| Malaysia[a] | 1982 | 640.8 | .. | .. | 1 330.0 | 521.0 | 100.0 | 2 591.8 | 0.7 |
| Sabah[a] | 1981 | .. | .. | .. | .. | 154.6 | .. | 154.6 | .. |
| | 1982 | .. | .. | .. | .. | 154.6 | .. | 154.6 | .. |
| Sarawak | 1981 | .. | .. | .. | .. | 139.4 | 11.7 | 151.1 | 2.0 |
| | 1982 | .. | .. | .. | .. | 144.1 | 39.5 | 183.6 | 21.5 |
| Maldives | 1981 | .. | .. | .. | .. | 2.3 | .. | 2.3 | .. |
| | 1982 | .. | .. | .. | .. | 2.3 | .. | 2.3 | .. |
| Nepal[a] | 1981 | 53.0 | .. | .. | .. | 25.0 | .. | 78.0 | .. |
| | 1982 | 113.0 | .. | .. | .. | 25.0 | .. | 138.0 | 76.9 |
| New Zealand | 1981 | 3 993.9 | .. | 140.0 | 1 000.0 | 3.8 | 689.0 | 5 826.7 | .. |
| | 1982 | 3 992.2 | .. | 135.0 | 1 000.0 | 3.5 | 689.0 | 5 819.7 | (0.1) |
| Pakistan | 1981 | 1 847.0 | 125.0 | .. | 1 484.0 | 15.0 | 581.0 | 4 052.0 | 21.5 |
| | 1982 | 2 547.0 | 125.0 | .. | 1 484.0 | 15.0 | 581.0 | 4 752.0 | 17.8 |
| Papua New Guinea | 1981 | 99.8 | .. | .. | 148.0 | 70.0 | 20.0 | 337.8 | 3.0 |
| | 1982 | 100.6 | .. | .. | 218.0 | 70.0 | 40.0 | 428.6 | 0.5 |
| Philippines[b] | 1981 | 939.8 | .. | 501.0 | 2 155.0 | 307.2 | .. | 3 903.0 | .. |
| | 1982 | 1 266.8 | .. | 559.0 | 2 155.0 | 343.2 | .. | 4 324.0 | .. |

**Table 1.5.** *(continued)*

| | | Installed generating capacity (MW) | | | | | | | *Increase over previous year (percentage)* |
|---|---|---|---|---|---|---|---|---|---|
| | | *Hydroelectric* | *Nuclear* | *Geothermal* | *Steam* | *Diesel* | *Gas turbines* | *Total* | |
| Republic of Korea | 1981 | 1 202.0 | 587.0 | .. | 8 065.0 | 1 026.0 | 208.0 | 11 088.0 | 7.2 |
| | 1982 | 1 202.0 | 1 266.0 | .. | 8 015.0 | 1 026.0 | 88.0 | 11 597.0 | 4.6 |
| Samoa | 1981 | 1.2 | .. | .. | 2.5 | 10.1 | .. | 13.8 | .. |
| | 1982 | 4.6 | .. | .. | 2.6 | 10.2 | .. | 17.3 | 26.0 |
| Singapore | 1981 | .. | .. | .. | 1 950.0 | .. | 60.0 | 2 010.0 | .. |
| | 1982 | .. | .. | .. | 1 950.0 | .. | 156.0 | 2 106.0 | 4.8 |
| Sri Lanka | 1981 | 372.3 | .. | .. | 50.0 | 20.0 | 80.0 | 522.3 | 23.7 |
| | 1982 | 372.3 | .. | .. | 50.0 | 20.0 | 120.0 | 562.3 | 7.7 |
| Thailand | 1981 | 1 361.1 | .. | .. | 2 280.5 | 350.4 | 610.0 | 4 601.9 | 14.8 |
| | 1982 | 1 519.9 | .. | .. | 2 329.5 | 358.3 | 850.0 | 5 057.7 | 9.9 |
| Viet Nam | 1981 | 280.0 | .. | .. | 510.0 | 360.0 | .. | 1 150.0 | .. |
| | 1982 | 280.0 | .. | .. | 510.0 | 370.0 | .. | 1 160.0 | .. |

*Source:*   Electric Power in Asia and the Pacific 1981 and 1982 (ESCAP).

a   Preliminary figures not confirmed.

b   Data relating to the National Power Corporation only.

*Note:*   No data available for Brunei Darussalam, Democratic Kampuchea, Guam, Lao People's Democratic Republic, Nauru, Niue, Solomon Islands, Tonga, Trust Territory of the Pacific Islands, Tuvalu or Vanuatu.

## 2. NUCLEAR RAW MATERIALS

Uranium and thorium resources in the Asian and Pacific region are shown in tables 2.1 and 2.2. There are major resources of uranium in Australia. Elsewhere in Asia, minor uranium resources might possibly be located in Indonesia, Pakistan, the Philippines, the Republic of Korea and Thailand. Small amounts of uranium of the order of a few hundred tons a year have been mined in India and Japan. China is believed to have relatively large, reasonably assured resources and also estimated additional resources, but details of these are not available.

Recent and anticipated trends in the production of uranium, thorium and monazite in the ESCAP region are shown in tables 2.3 and 2.4. An extensive review has been published on the world resources, production, supply and demand and exploration of uranium including information from national reports on uranium exploration, resources and production, and including national reports on thorium resources.[5]

From these tables, it can be observed that significant reserves exist in the ESCAP region. For the estimated probable 42 to 45 GW nuclear electricity capacity by the turn of the century, the region would require about 6 to 7

thousand tons of uranium every year at a steady, assured rate (assuming 33 per cent cycle efficiency and 95 to 100 per cent capacity). Although the attainable production (table 2.3) can meet the requirement, supply may be limited by the uranium enrichment facilities attainable in the region at the early stages. Thus, in the early years, the region may have to rely on imports. Similarly since very limited reprocessing capacity exists in the region, a reliable supply of enriched uranium fuel may pose a problem to the development of nuclear electricity.

The data on annual production of uranium in China are not available. At present only a few hundred tons a year of uranium are mined in Japan. During 1983-1984 India continued its integrated surveys and exploration for building up uranium and other nuclear raw materials to support the country's nuclear power programme utilizing modern techniques. New indications were located and are being investigated. Drilling was carried out at 17 different prospects in various parts of the country. Total indicated and inferred resources of about 73,000 tons of uranium oxide have been established. To ascertain the reserves of monazite, ilmenite and other heavy minerals, large areas in Kerala were investigated and the drilling of 717 holes was completed during the period.

The Republic of Korea has some low-grade uranium resources. About 68 million tons of uranium ore reserves

---

[5] "Uranium" December 1983 : a Joint Report by the Nuclear Energy Agency and IAEA, published by the Organisation for Economic Co-operation and Development (OECD).

---

### Table 2.1. Uranium resources in the ESCAP region and the world
(In thousands of tons)

| Country or area | Recovery costs for reasonably assured resources | |
| --- | --- | --- |
| | Up to $80/ kilogram | $80-$130/ kilogram |
| Australia | 294.0 | 23.0 |
| India | 31.0 | 10.7 |
| Japan | 7.7 | – |
| Philippines | .. | – |
| Republic of Korea | .. | – |
| World[a] (rounded) | 1 733.0 | 659.0 |

*Source:* Based on World Energy Conference, "Survey of Energy Resources 1983".

[a]     Excluding centrally planned economy areas.

*Note:*    .. Data not available.

## Table 2.2. Thorium resources in the ESCAP region and the world
### (In thousands of tons)

| Country or area | Recovery costs for | | | | Source |
| | Reasonably assured resources | | Estimated additional resources | | |
| | Lower cost up to $75/kilograms(kg) Th | Higher cost $75/kilograms(kg) Th | Lower cost up to $75/kilograms(kg) Th | Higher cost $75/kilograms(kg) Th | |
|---|---|---|---|---|---|
| Australia[a] | – | 17.6 | – | 0.2 | |
| India | 319.0 | 0 | 0 | 0 | WEC (1979) |
| Iran | 0 | 0 | 30.0 | 0 | NEA/IAEA (1979) |
| Malaysia | 18.3 | .. | .. | .. | Resource data quoted in WEC, "Nuclear Resources". |
| ESCAP region[b] | 337.3 | 17.6 | 30.0 | 0.2 | |
| World[b] (rounded) | 725 | 490 | 860 | 1 240 | |

*Source:* Based on World Energy Conference, "Survey of Energy Resources 1980".

[a] Since no cost category made available, resources are allocated to higher cost category.

[b] Excluding centrally planned economy countries (data not available).

*Note:* .. Data not available.

WEC = World Energy Conference
NEA = Nuclear Energy Agency
IAEA = International Atomic Energy Agency

## Table 2.3. Uranium production to 1981
### (In thousands of tons)

| Country or area | Production | |
| | Cumulative end of 1981 | In 1981 |
|---|---|---|
| Australia | 14.14 | 2.85 |
| India | .. | .. |
| Japan | 0.05 | .. |
| ESCAP region[a] | 14.19 | 2.85 |
| World[a] (rounded) | 487.00 | 43.90 |

*Source:* Based on World Energy Conference, "Survey of Energy Resources 1983".

[a] Excluding centrally planned economy areas (data not available).

*Note:* .. Data not available.

**Table 2.4.  Recent monazite and thorium production
in the ESCAP region and the World**
(In thousands of tons)

| Country or area | Monazite | Thorium Production | Thorium Capability |
|---|---|---|---|
| Australia | 4.27[a] | 0.78[b] | 1.0[b] |
| India | 2.99[c] | 0.16[c] | 0.36[c] |
| Malaysia | 1.72[c] – 2.00 | 0.11[c] | 0.18[c] |
| Sri Lanka | 0.01[c] | .. | .. |
| Thailand | 0.37[c] | 0.01[c] | 0.05[c] |
| ESCAP region[d] | 9.36 – 9.64 | [e] | 1.59 |
| World[d] (rounded) | more than 11 | [e] | more than 4 |

*Source:  Op. cit.*

[a] *Source:*  United States Bureau of Mines, *Mineral Year-book* (1975).

[b] *Source:*  World Energy Conference *Survey,* table 7, part I, Uranium and thorium: estimated production in 1978 (1979).

[c] *Source:*  World Energy Conference, "Nuclear re-sources", table D-2; World monazite and thorium production (1975).

[d] Excluding centrally planned economy areas (data not available).

[e] Totals are meaningless because figures apply to different years; the INFCE study indicates total world thorium production is of the order of 150 tons per year.

with an average grade of 0.04 per cent $U_3O_8$ (a mixture of uranium oxides) have been found through intensive surveys since 1970.

Possibly small deposits might be found in Indonesia, Pakistan, the Philippines and Thailand.

While the above regional survey of resources is of interest for the countries in the region, discussions at IAEA lead one to believe that the availability of uranium resources is not a constraint for the region in the medium term due to global overproduction.

Part Two

APPLICATIONS OF RADIOISOTOPES AND RADIATION

# 3. AGRICULTURE : FOOD AND FOOD PRESERVATION

The ESCAP region embraces 55 per cent of mankind, more than half of the earth's surface and all possible agro-climates. The region offers suitable agro-ecological conditions for the cultivation of practically all the economic plant species. Some agricultural indicators of this region are given in table 3.1. Of the countries in the ESCAP region, seven fall in the least developed category, and six are highly developed in terms of indices of farm productivity and the use of superior agricultural technology. For the rest, low agricultural yields, and poor capital and technical inputs are common denominators.

A priority area concerning the population in the ESCAP region is the availability of food and nourishment for the 2,600 million people living in Asia and the Pacific. Nutritional deprivation for millions of children owing to the lack of calories for subsistence, let alone a balanced diet, seriously debilitates normal, physical and mental development. Short-term solutions are urgently needed to decrease population growth in the region. Increasing food production and its availability, equity in distribution, increasing purchasing power, minimizing pre- and post-harvest losses, promoting better storage and longer shelf-life of food are important short-term solutions to the problems.

In some developing countries, only about forty per cent of the potential land is used. Further expansion of the arable area is needed. The "Green Revolution", the symbol of international co-operation in agriculture in the last two decades dramatically increased food grain production. More than seventy per cent of this resulted from higher yields rather than from an increase in the area under cultivation. An increase in productivity of arable land can be achieved by increasing the number of crops grown each year and increasing the yields of crops harvested. Increased production in developing countries requires increased use of fertilizers which are affected by the rising prices of fossil fuels. Problems of soil degradation occur with the increased use of external inputs. Research and studies are needed to conserve the land to enable its intensive use. To make plants be more productive, research is essential in the areas of increased photosynthetic efficiency, improved weeds control and bioengineering to adjust plant types to maximize productive potentials. Large-scale improvement in technology is needed if production targets are to be reached. Efforts must be mobilized to increase agricultural production without consuming too much expensive energy, and there must be great concern for the environment.

Isotopes and radiation techniques supplement other conventional methods in solving particular problems applied to food and agriculture. Nuclear techniques provide the most effective and sometimes the only means of addressing specific questions. In view of present scarcity or high cost of many agricultural commodities, nuclear techniques can indicate the most effective utilization of fertilizers, help to ensure that pesticide applications result in minimal ecological damage and help to improve crops for higher yields, better quality and disease resistance, and improve animal productivity increasing the output of high quality protein.

Since 1974 the two international organizations within the United Nations system, the Food and Agricultural Organization of the United Nations (FAO) and the International Atomic Energy Agency (IAEA), have established a joint programme for the specific purpose of assisting member States in applying nuclear techniques to the development of their food and agriculture.

In the ESCAP region, apart from the support of many research projects and co-ordinated programmes in many developing countries, the Joint FAO/IAEA Division had been responsible for the technical management of two large-scale projects in nuclear agriculture financed by the Swedish International Development Authority, in India and Bangladesh. In the region, Australia, Bangladesh, China, India, Indonesia, Japan, the Republic of Korea and Pakistan have strong programmes. Burma, the Islamic Republic of Iran, Malaysia, Mongolia, the Philippines, Sri Lanka and Thailand have modest programmes. Eight of the member States and all of the nine Associate Members of ESCAP do not have any programme in nuclear agriculture.

## 3.1 Mutation breeding

The purpose of crop improvement through mutation breeding is to produce varieties with desirable characteristics and higher yields. Ionizing radiation has been found to produce genetic variability at a much faster rate than nature. Artificially-induced mutants have been produced in several crops through this technique. Mutation breeding is established as a potent tool in the hands of breeders of grains, legumes, fodder, vegetables, fruits, sugar crops, industrial crops and ornamental plants.

The success of mutation as a viable breeding technique is evident from the expansion of such programmes; increasing numbers of crop species have been taken up for induction and institutes have become involved in mutation breeding. In the ESCAP region induced mutation has been

## Table 3.1. Agricultural indicators of some countries in the ESCAP region

| Country | Total land area (Thousands of hectare) | Arable land as percentage of total | Population | | Annual cereal production in thousands of tons | Calories/ caput/day | Protein/caput/ day (gram) |
|---|---|---|---|---|---|---|---|
| | | | Total | Agricultural (Percentage) | | | |
| **Developing** | | | | | | | |
| Bangladesh | 13 391 | 68.3 | 88 164 | 83.8 | 20 981 | 1 877 − | 43.9 |
| Bhutan | 4 700 | 2.0 | 1 296 | 93.4 | 104 | 2 028 − | − |
| Burma | 65 774 | 15.2 | 35 289 | 51.8 | 12 956 | 2 286 + | 55.9 |
| China | 930 496 | 10.7 | 994 913 | 59.8 | 286 018 | 2 472 + | 55.7 |
| Fiji | 1 827 | 12.9 | 630 | 40.2 | 23 | 2 628 + | − |
| India | 297 319 | 56.9 | 684 460 | 63.2 | 138 775 | 1 998 − | 49.5 |
| Indonesia | 181 135 | 10.8 | 148 033 | 58.9 | 33 558 | 2 295 + | 40.7 |
| Democratic People's Republic of Korea | 12 041 | 18.6 | 17 892 | 45.9 | 8 483 | 2 837 + | − |
| Republic of Korea | 9 819 | 22.4 | 38 455 | 38.6 | 8 005 | 2 996 + | 67.0 |
| Lao People's Democratic Republic | 23 080 | 3.8 | 3 721 | 73.7 | 1 056 | 1 856 − | 56.3 |
| Malaysia | 32 855 | 13.1 | 14 068 | 46.8 | 2 146 | 2 650 + | 51.4 |
| Maldives | 30 | 10.0 | 154 | − | − | 1 781 − | 58.2 |
| Mongolia | 156 500 | 0.8 | 1 669 | 49.0 | 302 | 2 711 + | 95.2 |
| Nepal | 13 680 | 17.0 | 14 288 | 92.6 | 3 575 | 1 914 − | 49.4 |
| Pakistan | 77 872 | 26.1 | 86 899 | 53.5 | 17 138 | 2 300 + | 58.4 |
| | (79 610) | (25.8) | (84 253) | (51.4) | (17 491) | (2 255 +) | (61.9)* |
| Papua New Guinea | 45 171 | 0.8 | 3 154 | 82.3 | 7 | 2 286 + | 45.1 |
| Philippines | 29 817 | 33.3 | 49 211 | 45.8 | 10 841 | 2 315 + | 47.7 |
| Samoa | 285 | 42.8 | 157 | − | − | 2 289 + | 48.5 |
| Sri Lanka | 6 479 | 33.2 | 19 815 | 53.2 | 2 063 | 2 249 − | 46.7 |
| Thailand | 51 177 | 35.1 | 47 063 | 75.4 | 21 092 | 2 301 + | 47.5 |
| Tonga | 67 | 79.1 | 97 | − | − | 3 221 + | 42.1 |
| Viet Nam | 32 536 | 18.6 | 53 740 | 70.6 | 12 183 | 2 029 − | 52.7 |
| Subtotal | 2 003 698 | 19.0 | 2 304 915 | 61.3 | 580 399 | − | 51.2 |
| **Developed** | | | | | | | |
| Australia | 761 793 | 5.8 | 837 | 5.8 | 21 003 | 3 202 + | 101.0 |
| Japan | 37 103 | 13.2 | 12 631 | 10.8 | 14 333 | 2 916 + | 82.4 |
| New Zealand | 26 867 | 1.7 | 288 | 9.3 | 821 | 3 511 + | 112.2 |
| Subtotal | 825 763 | 6.0 | 139 370 | 10.2 | 36 157 | | 85.1 |
| **Asia and the Pacific** | | | | | | | |
| Total | 2 829 461 | 15.2 | 2 439 285 | 58.5 | 616 556 | | |
| Percentage | 21.6% | 29.6% | 55% | | 38.7% | | |
| Rest of the world | 10 245 787 | 10.0 | 1 997 455 | 31.0 | 976 198 | | |
| World | 13 075 248 | 11.1 | 4 436 740 | 46.1 | 1 592 754 | | |

Data condensed from FAO-RAPA Monograph 2 (1983), Food and Agriculture Organization of the United Nations Regional Office for Asia and the Pacific, Bangkok.

\*   From Agricultural Statistics of Pakistan (1983).

used much more widely than in other parts of the world. More than sixty per cent of documented radiation-induced commercial mutant varieties in the world covering all categories of variability have originated in this region. Australia, Bangladesh, Burma, China, India, Indonesia, Japan, Pakistan, the Philippines, the Republic of Korea and Thailand are among countries which have already released many mutant crop varieties.

Apart from using mutant strains directly as commercial varieties, their use as sources of specific traits in conventional breeding programmes is gaining popularity.

About 600 varieties have been released through induced mutations; over 80 per cent of these cultivars have been evolved using radiation treatments.[1] Nearly 20 per cent were developed with chemical mutagens or combined treatments with various mutagens including radiation. Among these, 336 improved cultivars of cereals, other grain crops, vegetables, forage crops, fruits and industrial crops and more than 250 of ornamentals have been released for cultivation in more than 33 countries. Of the various mutant varieties, more than 30 have been bred through radiation-induced mutation in vegetatively propagated species excluding ornamentals.

Two examples of successful introduction of mutant varieties in rice are the mutant of IR-5 variety released as Shwe-war-Htun in Burma,[2] and the varieties based on an induced semi-dwarf mutant released in California. Shwe-war-Htun was reported to show good adaptation and possess high yield and superior grain quality. In 1980-1981 this variety was grown in an area of over 600,000 hectares. In 1980, more than 70 per cent of rice area in California was under induced semi-dwarf varieties.

In China, over 160 new varieties of grain, oil-bearing crops, fibre crops, vegetables, fruit trees, mulberry trees etc. have been developed through radiation breeding. The new crop varieties covered 6.7 million hectares of land in 1981, yielding 2.5 million tons of grain, 0.6 million tons of ginned cotton and 0.2 million tons of oil-bearing seeds.[3] China established a "Mutant Bank" in Beijing at the Institute for Application of Atomic Energy of the Chinese Academy of Agricultural Sciences.

---

[1] A. Micke, IAEA Bulletin, Vol. 26, 1984, p. 26-28.

[2] U.T. Myint, Proc. Agric. Res. Congr., Rangoon, Burma, 1981, p. 3-9.

[3] S. Guang-chang, New Frontiers in Technology Application, Tycooly International Publishing Ltd., Dublin, Ireland, 1983, p. 151-156.

## 3.2 Plant nutrients and soil water management

Tropical and subtropical rice lands are dominant in the ESCAP region. Other land types include irrigated semi-arid and arid lands, lands under shifting cultivation (e.g. in Burma and India), rain-fed and dry farming lands (e.g. in India and Pakistan). Each of these areas are characterized by specific problems and potentials. Progress made in food production is the outcome of expansion of irrigation, intensive use of fertilizers and input-responsive varieties of crops. In the Asian region, the low resource base of farmers limits the use of such high energy inputs. In developing countries about 40 per cent of potentially arable land is used and there are opportunities for expansion (see table 3.2.1). However, the expansion varies among regions and countries. It is expected that for the next two decades the increase in agricultural production will be the outcome of increase in yield (about 65 per cent), arable land (25 per cent) and from cropping intensity (10 per cent). Important areas of research to enable more intensive use of good quality land are related to optimal plant nutrient systems, soil moisture management and developing cultivation techniques with minimum commercial energy requirements.

The measurement of fertilizer uptake efficiency is carried out by using fertilizers labelled with isotopes, e.g. phosphorus, sulphur, calcium, zinc, iron, manganese and several other nutrients. At present the field-plot scale studies using isotopically labelled nitrogen fertilizer can be performed. Isotopes provide the only direct method for assessing fertilizer uptake. Results of many studies show that cereal crops take up fertilizer containing phosphorus more efficiently when the fertilizer is placed near the seed at planting than when it is broadcast and incorporated uniformly throughout. The results of the most effective placement of phosphorus fertilizer for flooded rice have changed fertilizer practices in large parts of the world. The economic returns from the exploitation of these results are massive. One country which participated in a research programme on nitrogen fertilization of maize estimated that it benefited about $US 36 million a year after its farmers had adopted new fertilizer placement practices. More efficient practices for fertilization of coconuts have been demonstrated in Sri Lanka. Additional savings are possible and it has been estimated that perhaps 50 per cent of the fertilizers used at present could be saved by improved practices based on sound research results. The results of many of these isotope aided studies were recently summarized by the Joint FAO/IAEA Division and published by FAO for use by agricultural industry specialists.[4]

---

[4] "Maximizing the efficiency of fertilizer use by grain crops", Fertilizer Bulletin, No. 3 (FAO, Rome, 1980).

**Table 3.2.1. Land under irrigation and arable land areas in selected Asian countries, 1969-1980.**
(Thousands of hectares)

| | 1969-1971 | | | 1980 | | |
|---|---|---|---|---|---|---|
| | Land under irrigation | Arable land area | Percentage | Land under irrigation | Arable land area | Percentage |
| Afghanistan | 2 340 | 7 877 | 29.7 | 2 680 | 7 910 | 32.9 |
| Bangladesh | 1 054 | 8 881 | 11.9 | 1 620 | 8 928 | 18.1 |
| Burma | 849 | 9 963 | 8.5 | 999 | 9 573 | 10.4 |
| China | 41 000 | 101 563 | 40.0 | 46 000 | 98 430 | 46.7 |
| India | 30 183 | 160 463 | 18.8 | 39 350 | 165 200 | 23.8 |
| Indonesia | 4 371 | 12 967 | 33.7 | 5 418 | 14 200 | 38.2 |
| Iran (Islamic Republic of) | 5 184 | 15 150 | 34.2 | 5 900 | 15 330 | 38.5 |
| Malaysia | 243 | 920 | 26.4 | 370 | 1 000 | 37.0 |
| Nepal | 116 | 1 923 | 6.0 | 230 | 2 316 | 9.9 |
| Pakistan | 12 904 | 19 119 | 67.5 | 14 300 | 20 030 | 71.4 |
| Philippines | 830 | 7 164 | 11.6 | 1 300 | 7 050 | 18.4 |
| Republic of Korea | 993 | 2 148 | 46.2 | 1 150 | 2 060 | 55.8 |
| Sri Lanka | 436 | 895 | 48.7 | 525 | 1 025 | 51.2 |
| Thailand | 1 965 | 12 244 | 16.0 | 2 650 | 16 250 | 16.3 |
| Viet Nam | 980 | 5 240 | 18.7 | 1 700 | 5 595 | 30.4 |

*Source:* Food and Agriculture Organization of the United Nations, *Production Yearbook 1981* (Rome, 1982).

On soil-water management water is often scarce for agriculture and this severely limits crop yields. Increased food production will therefore be linked in many cases with provision of water for irrigation. Increasing needs for crop production for growing population are causing the rapid expansion of irrigation throughout the world. As depicted in Table 3.2.1, in the ESCAP region China and India have the largest areas under irrigation. It has been considered that from its existing 40 million hectares alone, India should be capable of producing enough food to sustain twice the size of the expected population of 1 billion people by the year 2000. India plans to raise its irrigation potential to 60 million hectares by the year 1990. Bangladesh also plans to raise its irrigation potential from 15 to 25 per cent by 1990.

To increase food production it is essential to develop adequate water-management methods which would lead to better use of rainfall under dry farming conditions and to improve efficiency of water used on irrigated land. To achieve the target it is imperative to have a better understanding of the soil-water-plant-atmosphere continuum.

The fundamental principles of processes that govern the reactions of water and solutes within soil profiles are generally well understood. Soil-water profile studies give characteristics of the water economy of a field, of the quality of soil solution within the profile, and of water which leaches below the reach of plant roots into ground waters. Thus the results of studies will eventually lead to increased crop yield, to reduced loss of nutrients through leaching below the rooting zone and to the conservation of soil through avoidance of the accumulation of salts close to the soil surface. Under rain-fed agriculture, research results would help to control an erosion, conserve water and ensure sustained production at an acceptable yield level.

Isotopes are used as tracers to identify water movement. Studies in isotope hydrology involve naturally occuring isotopes and artificially applied isotopes. The basis for comparing various water management practices is normally provided by a very large number of soil moisture measurements in soil profiles. This had been difficult, costly and time consuming. Nuclear techniques involve the use of applied isotopes and the use of a neutron moisture probe. The latter makes it possible to follow the moisture changes in the soil profile in a reliable nondestructive manner. The isotope rubidium-86 had often been used to study the distribution of irrigation water under field conditions. In India, the studies made by the Indian Agricultural Research Institute (IARI) of soils of its farm indicated that the available moisture capacity varied from 7.5 to 12 centimetres (cm) for 60 cm and 11 to 20 cm for 100 cm profile depths. This indicated that they could not hold all the available moisture during rainy season and hence the root zone soil moisture reservoir was, in many cases, replenished in the early rainy season and subsequently the rainfall was disposed of as run off and the excess standing water caused waterlogging in the active kharif crop growth period. Field investigations conducted at a non-irrigated farm of IARI had shown that 30 to 60 per cent of the monsoon rainwater recharged the deeper layers of the soil profile. The information is useful in crop planning for dryland agriculture.[5]

## 3.3 Pests control and pesticide residues

It has been estimated that for the world's main food crops, the field losses from pests have been on the average of 35 per cent. In some places losses are higher. Pesticides are chemical compounds which are used to control pests such as insects, weeds, plant diseases, nematodes and rodents. It was estimated that pesticides have accounted for 20 per cent increase in the farm outputs in the United States since 1945. In the ESCAP region, crop losses caused by various pests may account for a substantial proportion of the potential harvest. The loss estimates for India are summarized in table 3.3.1. As cropping intensity increases the pest control measures would have to be increased correspondingly. The parallel relationship between the grain yield increase and corresponding increase in pesticides use, in India, is shown in figure 3.3.1. In the ESCAP region

### Table 3.3.1. Avoidable losses in Indian agriculture due to pests

| Type of pest | Percentage of total loss | Financial loss in 100 million Rupees |
|---|---|---|
| Weeds | 33 | 198 |
| Plant diseases | 26 | 156 |
| Insect pests | 20 | 120 |
| Miscellaneous pests | 8 | 48 |
| Storage pests | 7 | 42 |
| Rodents | 6 | 36 |
| Annual loss | | 600 |

*Source:* A.K. Banerji, *Pesticides*, 3, 1981.

---

[5] Unpublished report of the Indian Agricultural Research Institute, New Delhi, India.

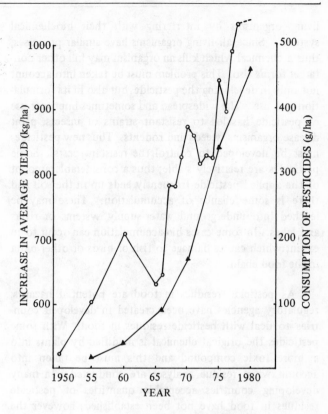

**Figure 3.3.1. Grain yield (o) increase in relation to pesticide use (△) in India (Reproduced from Mukerjee, 1982, IAEA-SM-263/35)**

one could group countries on the basis of utilization of pesticides per hectare of arable land and land under permanent crops in agricultural development as follows: Group I: 0-100 grams, Afghanistan, Bangladesh, Burma and Nepal; Group II: 100-500 grams, India, Indonesia, Pakistan and Papua New Guinea; Group III: 500-1,000 grams, Philippines, Sri Lanka and Thailand; Group IV: 1,000-5,000 grams, Malaysia and the Republic of Korea.

Competition from weeds for limited supplies of soil, moisture and plant nutrients results in losses to cultivated crop. Crop reductions of 50 to 70 per cent as the result of competition by weeds are not uncommon. Globally, more herbicides are used than insecticides. Weed control is of great importance to developing countries but herbicides are not used as extensively as in developed countries.

Control of vector-borne diseases of man and animals is based to a large extent on insect control since there are few vaccines available; thus insecticides play a major role in the control of malaria, filariasis, dengue and many other vector-borne diseases.

Protection of man, livestock and crops from weeds and insects by pesticides was found useful but the use of pesticides is not without problems. Many pesticides are toxic to living organisms by interfering with their biochemical systems. Since all living organisms have similar processes, thus a chemical which kills an organism may kill other non-target forms also. This problem must be taken into account not only in developing the pesticide, but also in its formulation and use. The widespread and sometimes improper use of pesticide has led to resistant strains of insects, plant disease organisms, weeds and rodents. Thus new pesticides must be developed to control the resistant pests. Some pesticides are relatively stable; thus a considerable amount of the applied pesticide frequently ends up in the soil and there is some chance of accumulation. These may be leached into underground water-supply systems or rivers and lakes. In some cases bio-accumulation can occur to an extent which causes damage to fish or birds or other biota in the food chain.

As pesticide residues in food are potential hazards, regulatory agencies have been created in developed countries to deal with pesticide residues in food. With some pesticides the original chemical is modified by plants into a more toxic compound and this must be taken into account when residue analyses are conducted. In many developing countries acceptable quantities of pesticide residues in food have not been established; however the guidelines developed by FAO and WHO are generally followed.

Investigations on pesticide metabolism in plant and other organisms, its movement in soil, its amount in water, its amount at various steps of food processing and its residue in food involve analyses for the presence of very small quantities of pesticide often present in parts per million. Conventional procedures are elaborate. The determination could be facilitated by using nuclear technique employing isotope-labelled pesticides. Certain data can be based from studies carried out in developed countries; however it is often necessary to conduct research in different developing countries, using specific pesticides on specific crops under local climatic conditions which influence persistence of pesticides differently.

A recent approach to pest control, integrated pest management, is based on ecological principles which involve integrating a number of disciplinary pest control methodologies. A thorough knowledge of the biology of the target species is required, in particular, its dispersal, population density and dynamics as well as the ecology of the natural enemies of the pest. Studies of these parameters can be greatly enhanced by use of nuclear technique, such as labelling the insects using stable or radioisotopes.

Another use of ionizing radiation for insect control is to expose insects to lethal doses of radiation as in the case of stored-grain product insects. The target species is thus killed, and the radiation leaves no residue as would have been the case if insecticides were applied. Ionizing radiation can also be used to sterilize the pest insect, so that when males are released in large numbers into the wild population, females mating with these sterile males fail to produce offspring. The sterile insect technique (SIT) has been developed by the Insect and Pest Control Section of the Joint FAO/IAEA Division and applied in large-scale field programmes to eradicate the mediterranean fruit fly in Mexico. The use of this technique to control fruit flies in Japan and in Taiwan Province has also proved successful and could be extended to other pests in the ESCAP region.

Different developing countries in the ESCAP region have been conducting studies on varying aspects of pest control including the fate of pesticide and pesticide residues by using isotope techniques and the sterile insect technique.

In future, it would appear that increases in food production will be attained through increased use of inputs including crop protection measures. The use of pesticides is increasing and will continue to increase for the remainder of this century as food production and population continue to increase. At the same time, there is the need to reduce to the minimum the residual effects of pest control measures in an attempt to maintain a clean environment. This need calls for more judicious use of pesticides and greater emphasis on an integrated approach to pest management.

## 3.4 Food preservation

After its harvest, agricultural produce is stored and processed in several ways before it is consumed, marketed or exported. Depending on the commodity and its vulnerability to adverse temperature, humidity and handling, there are several stages at which multiple losses occur. These losses could be caused by either mechanical, chemical, biological or other processes, yet they could be further aggravated by contamination with micro-organisms and parasites. The result is that gains made through better production are often nullified. The demand for prevention of losses is increasing in both developed and developing countries.

The extent of these losses varies from country to country and commodity to commodity (table 3.4.1). It has been estimated that 25 to 30 per cent of food produced in the ESCAP region is lost owing to spoilage. Losses could be as high as 35 per cent in the case of grain in some countries. In India, the loss of cereal grains, the major staple food, during storage has been estimated to be 10 to 11 million tons per year. Similar or bigger losses occur in more perishable food items such as fish, seafood, fruits, vegetables and other commodities. Contamination of food items with pathogenic organisms or parasites creates a public health problem leading to further indirect losses to the society.

**Table 3.4.1. Reported losses in some highly perishable commodities in the less developed countries.**

| Commodity | Production (Thousands of tons) | Estimated loss (Percentage) |
|---|---|---|
| **Roots/tubers** | | |
| Carrots | 557 | 44 |
| Potatoes | 26 909 | 5-40 |
| Sweet Potatoes | 17 630 | 35-95 |
| Yams | 20 000 | 10-60 |
| Cassava | 103 486 | 10-25 |
| **Vegetables** | | |
| Onions | 6 474 | 16-35 |
| Tomatoes | 12 755 | 5-50 |
| Plantain | 18 301 | 35-100 |
| Cabbage | 3 036 | 37 |
| Cauliflower | 916 | 49 |
| Lettuce | | 62 |
| **Fruits** | | |
| Banana | 36 898 | 20-80 |
| Papaya | 931 | 40-100 |
| Avocado | 1 020 | 43 |
| Peaches, apricots, nectarines | 1 831 | 28 |
| Citrus | 22 040 | 20-95 |
| Grapes | 12 720 | 27 |
| Raisins | 475 | 20-95 |
| Apples | 3 677 | 14 |

*Source:* National Academy of Sciences report, 1978 cited in *FAO Agricultural Service Bulletin* 43.

In earning foreign exchange, most of countries in the Asian and Pacific region could have manifold increases in the export of several specific food commodities. This would be possible only when the acceptability of these items in the world market increased through better preservation, sanitation and quality control. On importing food, developed countries have more stringent requirements and legislations. Quality and safety are two important considerations for world food commodity trade as much as for consumption by a country's own population. Therefore, storage technology for food, shelf-life and wholesomeness (safety for consumption) must be improved. These techniques have to be economical, least energy intensive, safe for consumers and handlers and above all acceptable to importers from the point of view of appearance and organoleptic properties.

The nuclear technique used in food preservation is radiation application. The process of food irradiation involves exposing food to ionizing radiation. Radiation sources are gamma-rays from cobalt-60 or caesium-137 or x-rays generated from machine sources operated at or below an energy level of 5 MeV; or electron beams generated from machine sources operated at or below an energy level of 10 MeV. Various dosimetry techniques are known for use in the measurement of radiation doses for different sources of radiation. Radiation levels in food irradiation vary according to purpose, for example, to reduce the number of viable organisms so that few if any are detectable; to reduce the number of viable, specific non-spore forming pathogenic micro-organisms (e.g. *salmonella*) so that none are detectable; to enhance keeping quality (shelf-life extension) by substantial reduction in number of viable specific spoilage micro-organisms; to interfere with physiological processes rather than microbiological ones, for example, sprout inhibition and delayed ripening of fruits. The energy levels used in food irradiation are too low to lead to any production of radioactivity in the irradiated food.

The three preconditions necessary for the widespread use of irradiated foods have been fulfilled for many food items. These are (a) proof of safety for human consumption (wholesomeness), (b) technical feasibility, that is it yields a product of acceptable or superior quality under practical conditions or it can save products from spoilage for a long time, and (c) economic competitiveness, that is the cost should not exceed its benefits.

Reports on technical feasibility are manifold; examples will be described subsequently on certain food items. The economic feasibility of any industrial operation can only be tested in large-scale experiments, that is using pilot-irradiators capable of handling from a few hundred to a few kilograms of material within a short period of time. The economic analysis of food irradiation has been shown to have a favourable cost: benefit ratio, for example 1:4 to 1:6 for papaya disinfestation; 1:5 for mushroom preservation. Depending on the size of operation (annual through-put capacity), the total operating costs of this new process vary from 0.9 to 3.2 per cent of the value of the product for sprout inhibition in potatoes and onions; from 0.19 to 2.4 per cent for grain disinfestation and from 0.2 to 2 per cent for the prolongation of the shelf life of fish.

The safety for human consumption (wholesomeness) of irradiated food is one of the most important aspects. The wholesomeness of irradiated food requires not only toxicological but also microbiological and nutritional approaches. The wholesomeness of food treated by heat, microwaves and other traditional methods has never been demanded. As the result of intensive wholesomeness studies in a number of laboratories of many countries, especially many advanced countries, during the past years, clearance or approval of several irradiated food commodities fit for human consumption has been granted by many national authorities.

Thirty years of development work on the preservation of food by irradiation have shown that the treated food is safe; that is, no harmful effect to animals and humans has been found and the method is efficient, that is it requires less energy than other preservation methods. The Joint FAO/IAEA/WHO Expert Committees on the Wholesomeness of Irradiated Food (1969, 1976 and 1980) have evaluated the safety for human consumption of irradiated foods and in 1980 concluded that the irradiation of any food up to an average dose of 10 kGy (kilo Gray: 10 kGy=1 Mrad) causes no toxicological hazard and hence toxicological testing of food so treated is no longer required.[6]

The Codex Alimentarius Commission has adopted a recommended international general standard for irradiated food.[7] The standard was based on the clearance of eight foods recommended by the 1976 Expert Committee on the Wholesomeness of Irradiated Foods. This was updated in the light of recommendations of the 1980 Expert Committee.

In 1979, the IAEA published guidelines for the Government of RCA countries to harmonize their national laws in accordance with the Codex Standard and the Code of Practice for the operation of their radiation facilities used for the treatment of foods. Incorporation of the regulations in existing national food laws would greatly facilitate international trade and ensure a similar and effective control over the irradiation of foods.

In the ESCAP region, for the past fifteen years about 12 countries have been engaged in studies of local food items for preservation by radiation utilizing small and large irradiators as may be available particularly in Australia, Bangladesh, China, India, Indonesia, Japan, Pakistan, Philippines, Republic of Korea, Malaysia, Singapore and Thailand. Recently additional multipurpose larger irradiators were either installed or being planned in Bangladesh, Indonesia and Thailand. Irradiated food items which have

been given clearance by national authorities are: frozen shrimp (1979), Australia; potato, onion, papaya, strawberry, chicken, rice, wheat, shrimp and frog-leg (1983), Bangladesh; potato (1972), Japan; potato (1972) the Philippines; and onion (1973), Thailand. There is a large food irradiator in Japan which irradiates potatoes on a commercial scale. The treated potatoes are used mainly to supply processing lines for the production of commodities such as potato chips and French fries during the off-season.

In developing countries in the ESCAP region, irradiated food items of interest are fruits, vegetables like onion and garlic, seafoods, cereal grains and spices. The ultimate purposes are similar, that is to prevent spoilage for domestic consumption as well as for exportation since each country is capable of producing more of these commodities.

Feasibilities for treating commodities with doses under 10 kGy and advantages of such radiation treatment are, for example:

*1 to 3 kGy:* This range of doses can extend the storage-life of iced fresh fish by a factor of between 2 to 5.

Doses of about 3 kGy will extend shelf-life by eliminating pathogens in cooked and dried shrimp. Irradiation could destroy certain pathogenic bacteria such as *Salmonella* and *Vibrio parahaemolyticus* in frozen seafood.

*0.075 to 0.75 kGy:* Radiation disinfestation of cereals, grains, beans and maize could be carried out by doses within this range.

On the other hand, mango irradiated with a dose of 0.75 kGy will give delayed ripening thereby making it possible to transport the fruit over long distances and simultaneously eliminate the plant quarantine problem by killing mango-weevil even in the seed of the fruit where no insecticide can possibly reach.

*0.05 to 0.15 kGy:* This range of doses would be effecto control sprouting in potatoes and onions.

The annual production of onions in India, Japan and the Republic of Korea is about 2.5, 1.2 and 0.25 million tons respectively. Most countries in the region can produce one crop a year. Results obtained in India show that irradiating onions reduced losses even under ambient condition storage while non-treated onions started sprouting after two months. When treated with a dose of 0.06 kGy and stored in a traditional storage shed little or no sprouting occurred. About 800 tons of onions treated with a dose of 0.1 kGy and stored in a cold storage at 10±2°C for six months were marketed successfully in Thailand during 1973-1974. No importation of onions would be needed in Thailand if the radiation treatment of the product

[6] "Wholesomeness of Irradiated Food", Report of a Joint FAO/ IAEA/WHO Expert Committee, WHO Technical Report Series, No. 659 (1981).

[7] Recommended international general standard for irradiated foods and recommended international codes of practice for the operation of irradiation facilities for the treatment of foods, Joint FAO/WHO Food Standards Programme, Codex Alimentarius Commission, CAC/RS 106-1979, CAC/RCP 19-1979.

is continued, and 40 per cent of the losses could be prevented annually.

*On spices:* Countries in Asia notably India, Indonesia and Malaysia are among the world's leading producers and exporters of spices (black pepper, white pepper, nutmeg, coriander, cassia bark and mace). Many developed countries are exporting more and more spices for use in processed food industries. Most spices from tropical countries are heavily contaminated with micro-organisms and insects. Such contamination often leads to the detention or condemnation of shipments by health authorities of developed countries. Pathogenic bacteria and toxigenic moulds such as *Aspergillus flavus* are found occasionally in imported spices. Heavily contaminated spices cause serious problems for the quality and value of processed food. Traditionally, spices are decontaminated by fumigating with ethylene oxide or propylene oxide. The effectiveness of treatment depends on temperature and humidity. Ethylene oxide, which is commonly used, is toxic to man and may react with constituents of spices to form compounds such as ethylene chlorhydrin and ethylene bromohydrin which persist as toxic residues. Studies in Indonesia show that a radiation dose of 5 kGy could reduce the microbial load of spices by a factor of between 100 and 10,000 and the volatile constituents of the product are not affected. Studies in Malaysia show that a dose of 9 kGy did not alter the composition of volatile components of black pepper. Slight effects were found on the treatment of white pepper but the effect stabilized during storage. The spice industries in Indonesia and Malaysia will co-operate with each other on irradiated spices and transport trials of irradiated spices from South-East Asian countries to Europe and the United States of America are planned. The cost of the multi-step fumigation process appears to be twice that of irradiation.

*On fruits:* Many varieties and large quantities of seasonal fruits are produced annually by most countries in the Asian and Pacific region. Most tropical fruits are sensitive to low temperature and under ambient conditions they cannot be kept for long. Each season there are great losses owing to spoilage. Local methods of preservation are usually used. The market for tropical fruits remains traditional with insufficient development utilizing modern technology. The demand for these exotic fruits abroad is great but approved quarantine treatments are required. The approved quarantine treatment of citrus, papaya and other fruits and vegetables consists of fumigation of the products with brominated organic fumigants notably ethylene dibromide (EDB). This has been widely practiced within countries and in international trade. Through active investigations abroad and in the region, it was determined that doses in the range of 0.075 to 1.0 kGy can give adequate quarantine on infested fruits, in some cases even when packaged in standard commercial containers.

The use of EDB is strongly condemned by national authorities in several countries. From 1 September 1984 any fruit containing EDB is not permitted for markets in the United States. The acceptable daily bromide intake for humans has been set at one part per million (ppm); 20 ppm and 0.1 ppm of bromide are permitted in raw cereals and in bread respectively.[8] Previously the rules for EDB were established on the basis of bromide toxicity, not on that of EDB itself. EDB can remain intact in food. There have been no guidelines for EDB on citrus fruit.

It appears that the irradiation technique could overcome the existing quarantine regulations for fruits. It gives no residue and requires a short period of treatment, thus reducing the interval between picking and shipment by as much as one day. Radiation treatment of finished packages reduces chances of reinfestation. Together with refrigerated storage, radiation treatment may also delay senescence of certain types of fruits, which could also reduce the pressure on air transport significantly. However, the commercial facility should be designed in such a way as to process commercial-sized loads with fairly uniform doses. Transport trials to determine the efficacy and economics of radiation treatment would have to be made.

The use of higher radiation doses (higher than 10 kGy) in the range of 20-45 kGy may be of interest for sterilization of meat, poultry and fish to produce shelf-stable foods, since it needs less energy than other treatments, such as a combination of heat and frozen storage. Meat slices treated by radiation and packaged in suitable plastic bags, had a shelf-life of several years at ambient temperature. Irradiation would reduce nitrite and nitrate requirements of cured ham by about 80 per cent, thereby reducing the cancer hazard associated with the nitrosamine content of hams produced by adding nitrate to the meat during curing. As a whole it has been considered that further information might be needed to suit nutritional, microbiological and toxicological implications, though a certain country or a few countries have given clearance to irradiated food items such as frozen shrimps, chicken and froglegs for domestic consumption.

Acceptance of irradiated food items for domestic consumption depends on national governmental regulatory agencies and consumers. Good manufacturing practices must be established so that retailers and consumers obtain high quality foods; otherwise objectionable quality may be associated with irradiation. A large-scale application of

---

[8] These figures might be changed by the Codex Alimentarius Commission in their recent studies.

food irradiation requires a commercially-organized agricultural system involving national authorities and the food industry in research and development projects. World-wide interest in food irradiation technology continues and continued efforts and support from international organizations, governments and the food industry will be needed for the introduction of food irradiation on a commercial scale.[9]

In 1972 the IAEA initiated for the region of Asia and the Pacific a co-ordinated research programme on food preservation. Later the RCA food irradiation project, initially on fish products studies, became a part of the IAEA programme. After joining RCA in 1978, Japan gave financial and technical support to the RCA food irradiation project and the range of food studies was extended to include mangoes, onions and spices. This so-called phase I of the regional project, concerning research and development including pilot-scale studies, ended in August 1984. Members of the project included Bangladesh, India, Indonesia, Japan, Malaysia, Pakistan, the Philippines, the Republic of Korea, Sri Lanka, Thailand and Viet Nam.

Phase II of this regional project will be supported by Australia for three years as from April 1985. The object of

[9]   J. van Kooij, IAEA Bulletin, Volume 23, No. 3 (September 1981).

co-operation in this phase is to transfer food irradiation technology to local industries for commercialization.

## 3.5 Animal production and health

Livestock has been and continues to be the backbone of traditional agriculture. With increased human population and cropping intensity the role of livestock is becoming more and more important in the integrated farming systems of developing countries in the ESCAP region. Livestock's contribution as an energy input into traditional farming is as important as its role as supplier of valuable proteins to human diet and organic material to the soil. Livestock also contributes other products to the national economy. In developing countries in the ESCAP region the largest proportion of the domestic livestock is the ruminant species including cattle, sheep and goats (see table 3.5.1) Animal production in developing countries is limited by poor reproductive performance, lack of efficient use of available feedstuff and poor adaptation of the animals to the environment and infectious diseases.

At the cost of improving the traditional, well-adapted and tropical-heat-tolerant breeds of cattle, there has been a trend in the Asian and Pacific region to introduce temperate cattle for higher productivity. This approach has not

Table 3.5.1. Livestock population of some countries in the ESCAP region (1980)
(Thousands of head)

| Country | Cattle | Buffalo | Sheep | Goats | Pigs | Poultry |
|---|---|---|---|---|---|---|
| Bangladesh | 21 590 | 512 | 516 | 9 266 | – | 67 000 |
| Burma | 7 702 | 1 803 | 217 | 577 | 2 279 | 21 000 |
| China | 64 600 | 30 000 | 102 880 | 80 262 | 319 705 | 800 000 |
| India | 181 992 | 60 698 | 40 700 | 70 580 | 9 410 | 144 000 |
| Indonesia | 6 534 | 2 506 | 4 197 | 7 906 | 3 296 | 146 000 |
| Pakistan | 15 083 | 11 546 | 26 239 | 30 203 | – | 58 000 |
| Philippines | 1 885 | 2 760 | 30 | 1 450 | 7 599 | 6 300 |
| Mongolia | 2 397 | – | 14 230 | 1 985 | 34 | 249 |
| Malaysia | 481 | 199 | 59 | 312 | 1 393 | – |
| Thailand | 5 000 | 6 250 | 61 | 31 | 5 547 | 80 000 |
| Viet Nam | 1 450 | 2 200 | 14 | 200 | 9 354 | 84 000 |
| Sri Lanka | 1 720 | 898 | 30 | 512 | 93 | 6 300 |
| Republic of Korea | 1 506 | – | 4 | 197 | 1 832 | 43 000 |
| Democratic People's Republic of Korea | 950 | – | 290 | 240 | 2 100 | – |
| Australia | 25 168 | – | 134 407 | – | 2 430 | – |
| New Zealand | 8 131 | – | 68 722 | – | 434 | – |
| Japan | 4 385 | – | 16 | 62 | 10 065 | 295 968 |

Based on Food and Agriculture Organization of the United Nations statistical survey.

been readily successful owing to the inability of the exotic cattle to withstand stress and tropical diseases. More research is needed to develop management and disease-control practices to minimize the effects of stress. Only recently has the attention shifted to some extent to domestic breeds and species.

Nuclear techniques now play an integral part in animal husbandry research to improve health and productivity of animals.

Diseases are important factors limiting animal production. Many practical problems could be solved by application of nuclear techniques. For example, worm infections can seriously reduce livestock productivity or even kill animals in large numbers. The normal methods of using anthelmintics may result in the development of resistance among parasites. Another, and more permanent approach to controlling parasitism is to vaccinate the animals. The nuclear technique is to expose infective larvae to ionizing radiation. This has the effect of rendering the parasite non-pathogenic but still able to stimulate the animal's immune defence system when ingested. As a result, when the animal is exposed to infection in the field, the larvae which are ingested are unable to develop and the animal remains healthy. Radiation-attenuated vaccines have been developed to protect farm animals from several parasitic diseases. In India these vaccines have now been accepted as a normal procedure in lungworm disease of sheep in the Indian hill States. Radio-attenuated vaccines hold a big promise against other diseases for large and small animal raisers in the Asian and Pacific region.

On animal nutrition, in many developing countries there are shortages of cereal grains even for human consumption and thus the use of these food sources to supplement animal diets cannot be realistically considered. Animals are usually left to feed on pasture lands, on low quality roughage materials or on agro-industrial by-products which are of little or no use to man. Ruminants with specialized microbial fermentation digestive systems in their multi-compartment stomachs are well suited to poor-quality feed. In applying isotope techniques materials labelled with C-14, H-3, S-35, N-15 and P-32 are used for assessing the products of the rumen particularly for volatile fatty acids which become an energy source and for microbial proteins which become a protein source for the animal. These tracers provide an accurate and reliable means of measurement in the complex system of the multi-stomached animals. Such measurement has not been possible by other means. By using nuclear techniques intensive research programmes to identify ways of obtaining more efficient utilization of locally available feed resources are being carried out. In developing countries in the ESCAP region rice-straw, bagasse, water-hyacinth, leaves and stems

of banana, etc. have potential as animal-feeds. It was found that all that may be needed is to supplement the poor diet with a non-protein nitrogen source such as urea, or other nutrient or mineral to make up for an imbalance or deficiency in some cases. In a study on nitrogen metabolism using N-15, microbial protein and rumen bypass-protein synthesis was demonstrated. A small energy- and protein-rich supplement was found to increase animal's appetites, giving the animal not only more total nutrients but also substantial increases in the protein from both dietary (rumen bypass) and microbial sources. Such supplement permitted a substantial growth by the animals. Studies conducted in Sri Lanka and Bangladesh indicated that urea-ensiled straw was a suitable source of roughage and increased milk production in milking cows. In studies on buffaloes in Indonesia using P-32 labelled materials, it was concluded that rice bran had the ability to increase the digestability of the native grass through promotion of microbial protein synthesis.

One of the principal ways to increase livestock productivity is to improve reproductive efficiency. To a large extent, this could be achieved by reducing the generation interval (i.e. the time from birth of the animal to that of its first offspring). This interval is known to be exceptionally long for livestock in many Asian countries. However, to reduce the generation interval requires thorough investigation of the nature and aetiology of the common reproductive problems encountered under the very different environment and management conditions existing in the region, and the subsequent formulation of programmes applicable at the field and small-farm level for alleviating them.

Radioimmunoassay techniques have given researchers the opportunity of measuring and following patterns of the hormones which control reproductive processes. These techniques require a blood or milk sample and do not involve any administration of radioactivity to the animal. To the test tube containing the sample is added antibody and antigen isotopically labelled with I-125 or H-3. The data subsequently provided can be used to differentiate non-pregnant from pregnant animals; to monitor the response of animals to corrective therapy; to monitor the onset of sexual maturity (puberty); to confirm oestrus; and in general, to detect sub-optimal ovarian and testicular function.

Several programmes on the use of nuclear techniques in animal science in the Asian and Pacific region have been supported by the Joint Division of FAO/IAEA (see table 3.5.2).

One of these programmes which was a part of RCA and was terminated in February 1984 focused on improving

**Table 3.5.2. FAO/IAEA co-ordinated projects in the fields of livestock in the ESCAP region.**

| Area | Project | Countries participating/participated |
|---|---|---|
| Nutrition | Isotope-aided studies on non-protein nitrogen and agro-industrial by-products utilization by ruminants. | Australia, Bangladesh, Malaysia, Republic of Korea. |
| Multi-dimensional | Use of nuclear techniques to improve domestic buffalo production in Asia: nutrition, reproduction, diseases, metabolism and adaptation. | Australia, Bangladesh India, Indonesia, Malaysia, Philippines, Sri Lanka, Thailand. |
| Health | Use of nuclear techniques in the study and control of parasitic diseases of farm animals. | Australia, Sri Lanka. |
| Reproduction | The application of radioimmunoassay to improve the reproductive efficiency and productivity of large ruminants. | Australia, Indonesia, Malaysia, Sri Lanka, Thailand. |
| Multi-dimensional | Improving sheep and goat productivity with the aid of nuclear techniques. | Australia, China, Malaysia, Sri Lanka, Thailand. |
| Health | The use of isotope techniques in research and control of ticks and tick-borne diseases. | Australia, India |
| Multi-dimensional | Use of nuclear techniques to improve domestic buffalo production in Asia, phase II. | |

domestic buffalo production. The results of this highly successful programme are now being prepared for publication. Researchers dealing with reproduction have characterized the reproductive capacity of different breeds of buffalo in different countries with a view to gaining an understanding of the causes of the long-recognized low reproductive capacity of these animals. Working radioimmunoassays for reproductive hormones have been established in India, Indonesia, Malaysia, the Philippines, Sri Lanka and Thailand. The data generated from the programme clearly point to the fact that radioimmunoassay has helped substantially to shed light on important aspects of reproduction in the water buffalo. Thus suckling has been identified as the major cause of the long calving interval in the swamp buffalo. Short-term calf removal seems, according to the research undertaken, to be one promising managerial practice to overcome this problem. In the area of nutrition, considerable amount of information has now been obtained on the utilization of straw-based diets for buffalo, and on the beneficial effects on digestibility of treatment with urea. Isotope-based studies on the patho-

genesis, immunology and control of toxocara infections have helped to define the nature and development of the disease, the host response and strategies for the control of the infection in calves.

Without doubt, this programme has already helped to increase buffalo productivity through improved management practices derived from research findings. It has also enhanced substantially the level of expertise as well as the educational quality within many university departments and research institutes in the RCA region; it has encouraged close contact between scientists within and outside the region; and it has promoted information exchange on an animal which is vital to the interests of the region's small farmers and agricultural industry.

IAEA has agreed to support a follow-up programme on buffaloes. The new programme has attracted considerable interest from scientists in South-East Asia. Currently, contract proposals for the new programme are being evaluated.

# 4. APPLICATIONS OF RADIONUCLIDES AND NUCLEAR RADIATION

## 4.1. Medical applications of radionuclides

When radioisotopes and their detecting and measuring equipment became widely available to medical researchers, the application of radioisotopes (radionuclides, radiopharmaceuticals) in medicine began to develop very quickly. In any developing country, the introduction of radioisotopes for use in medical fields has been more widespread than in agriculture and industry.

Recent developments have made radioisotopes an essential tool in any large hospital. These developments are: availability of many short-lived radiopharmaceuticals, of better detection equipment, application of radioimmunoassays, availability of more accurate and easier dosimetry methods and introduction of on-line computers.

In radiotherapy, x-ray equipment has been widely used in cancer diagnosis and therapy. As the result of more availability of cobalt-60 produced by the use of nuclear reactors, this radioisotope, which is cheaper than radium, has also been used as an external radiation source. The cobalt-60 teletherapy unit permits a number of doses to be given from different angles resulting in a build-up dose at the tumour. The radiation source can be mounted so that it can rotate about the patient's body. Cobalt-60 seeds, similar to radium needles, can be planted in the tumour. Also *in vivo* therapeutic technique is to locate a suitable short-lived radioisotope, which can be administered orally, as a radiation source in the diseased tissue, e.g. iodine-131 to concentrate in the thyroid and phosphorus-32 to treat a type of blood cancer.

The unique character of radionuclides is that they are easily traceable (as tracers) in small amount in the body, *in vivo*. It is medical diagnosis which benefits most from this fact. Improved instrumentation and technology from linear scanner to computed gamma-camera and positron computed tomography or positron camera etc. and newly added radiopharmaceuticals widen the range of application. With these devices, not only can spatial distribution of radionuclides in an organ be assessed but also the dynamic process taking place in the organ itself. By using a colour system, multiple functions and spatial relations can be viewed in the same image. If radionuclides emit positrons, three-dimensional images can be constructed. These sophisticated and expensive instruments are used in some developing countries in the ESCAP region in major government hospitals and, in some cases, in private clinics.

Many diagnostic studies can only be accomplished through the use of radiopharmaceuticals. They are applied in thyroid, hepatic, renal, cardiac, pulmonary, gastrointestinal, in regional blood flow studies, etc. The best known isotope is iodine-131 which has been used for quite a long time. The iodine-131 compounds are used to measure the function of the thyroid, the liver, and the kidney as well as to make blood volume and circulatory studies; others are chromium-51 for spleen screening, selenium-75 for pancreas investigations and cobalt-57 for pernicious anaemia diagnosis.

In order to reduce the already relatively small radiation burden for each *in vivo* investigation, long-lived radionuclides are getting replaced with short-lived ones. Short-lived isotopes are not economically transported over long distances but now few can be produced daily by using specially designed "isotope generators". That is the (parent) long-lived radioisotope gives a decayed product (daughter) which is the desired short-lived one, which has to be separated from the parent and then incorporated into the right chemical form. The daughter isotope decays in a matter of hours and gives less radiation to the patient, for example technitium-99m (half-life, 6 hours) produced from molybdenum-99 (half-life, 2.7 days) and indium-113m (half-life, 1.67 hours) produced from Sn-113 (half-life, 118 days). The costs of "isotope generators" are within reach of medical research workers. With some effort certain assemblies have been produced within developing countries in the region, for example India, Indonesia and the Republic of Korea.

The radioimmunoassay and related procedures for measurement of hormones, enzymes, vitamins, some drugs, certain serum proteins and a number of other substances in the body fluids, tissues and blood are now in the front rank of medical applications of radioisotopes. The techniques are highly specific and sensitive. Radioimmunoassay procedures are carried out on specimens in the laboratory and do not involve administration of any radioisotope to the patients for diagnostic tests. The radioimmunoassay of great potential for tropical countries is the radioimmunoassay of hepatitis antigen which enables mass screening of donor blood for the occurrence of the dangerous hepatitis virus. Because of the growing availability of reagents and kits, radioimmunoassays are now widely used in the routine diagnosis and investigation of diseases.

Among developing countries in the ESCAP region, kits for the *in vitro* radioassays e.g. radioimmunoassay of insulin, human placental lactogen, radioimmunoassay of T-3 (tri-iodo-L-thyronine) and T-4 (thyroxine) have been locally produced, for example, in India, for widespread domestic uses. During 1982, over 440,000 nuclear medical investigations have been carried out in India using radio-pharmaceuticals produced by the Bhabha Atomic Research Centre, Trombay. The Centre supplied solvent extraction based technitium-99m generators which are regularly used in 35-40 institutions in India. It is estimated that 350-400 curies of molybdenum-99 are being supplied annually.

As mentioned in section 3.5, radiation-attenuated vaccines have been developed to protect farm animals from diseases that adversely affect animal production. Recently a method was developed which promises to produce a vaccine against malaria. However major uncertainties still remain which necessitate additional research.

Among developing countries in the region, a few examples could be described showing increased uses of nuclear techniques in medical diagnosis and treatment. Pakistan has established eight nuclear medicine centres while one more is being established. In 1984 in Pakistan about 100,000 patients were investigated and treated at these eight centres. Bangladesh established a nuclear medicine centre in Dhaka in 1962. Two more centres were set up outside the capital city in 1970. In 1981, the Institute of Nuclear Medicine was established in Dhaka and in 1982 two more nuclear medicine centres started to function. Of these centres, all except one are attached to medical college hospitals. There were about 320 diagnostic and therapeutic X-ray machines in use in the country in 1982. Some scanners and a gamma-camera became available later for uses of radioisotopes in medical diagnosis. When a gamma-camera with computerized data and image processing system began to operate, about 5,000 patients were diagnosed in one year utilizing this equipment. About 25,000 patients received diagnosis utilizing radioisotope technique in all nuclear medicine centres during 1982. In the same year radioimmunoassay tests were carried out accounting for 17 per cent of patients diagnosed by radio-isotope technique.

In Thailand, for the period from 1978 to 1982, the number of diagnostic X-ray machines increased from 800 to 1,700 and that for the therapy machines from 57 to 65 while the Co-60 teletherapy units increased from 10 to 19. However, during the same period the radium sources (tubes, needles and plaques) amounted to the radiation activity of about 5,000 milligrams radium in each year.

Since 1978, there were 22 nuclear medicine units including four in medical research institutions and two in private hospitals. Out of 16 nuclear medicine units, six are in medical university hospitals of which three are in Bang-

kok and three are in provincial university hospitals. The remaining units are in government hospitals in the Bangkok metropolis. The users of radioisotopes and radioimmuno-assay technique in diagnosis and in therapy are mainly concentrated in Bangkok, particularly in the three medical university hospitals and two government hospitals. The three provincial medical university hospitals have moderate programmes; however the one in Chiangmai has been developing its programme very rapidly.

About 76,000 patients received diagnosis utilizing radioisotope techniques in 1978; however the average figure for 1979 to 1982 could be taken as about 100,000 patients per year[10].

In medical diagnosis the use of *in vitro* tests amounted to 15,000 in 1978 representing 37 per cent of the use of radioisotopes in diagnosis. The number increased to 45,000 tests in 1982. The technique accounted for 82 and 65 per cent of the use of radioisotope techniques in the nuclear medicine units of the two medical university hospitals in Bangkok. The use of radioisotopes in therapy was largely confined to the use of radium and Co-60 and the number of patients receiving treatment increased from 19,000 in 1978 to 47,700 in 1982.

## 4.2 Radiation metrology

Earlier on, it was noticed that radiation doses given to hospital patients were not quite correct because of the lack of available standards. It was realized that dosimetry requirements for radiotherapy centres should be established in order that patients (e.g. cancer patients) received the right amount of ionizing radiation.

Furthermore, in many developing countries nuclear activities have been expanding. Nuclear techniques are being used not only in medicine but also in other fields such as agriculture and industry. These activities require a reliable radiation protection service for radiation workers.

Reliable and accurate radiation dose measurements to assure radiation safety were demanded for the protection of man in all activities in which ionizing radiation is intentionally used, including radiotherapy and industrial processing and this led to the establishment of the IAEA/WHO international network of secondary standard dosimetry laboratories (SSDLs) in many developing countries. The

---

10  The data on the use of radioisotopes in diagnosis and treatment in government provincial, district and certain large private hospitals are not readily available.

purpose of these laboratories is to bridge the gap between primary measurement systems and the users of ionizing radiation, e.g. the hospitals. Such a laboratory must have properly calibrated radiation sources and reference dosimeters and must be linked to the world's dosimetry system in order to ensure calibration measurements against primary radiation standards.

The IAEA Dosimetry Laboratory at its Headquarters has become the established centre for intercomparisons, training and advice. Services in radiation dosimetry are still provided using pre-calibrated radiation dosimeters which are sent by post to participating laboratories. Since 1972, about 1500 Co-60 radiotherapy dose comparisons in 70 IAEA member States have been performed by IAEA and WHO. The doses now measured show a vast improvement in the limits of error, and by means of this standardized control many hundreds of patients in many countries have benefited from the establishment of SSDLs.

The network is scientifically supported by 12 national primary standards laboratories and 5 collaborating international organizations including the International Bureau of Weights and Measures.

SSDLs in the ESCAP region are located at (the figure in the bracket is the year designated):

(1977)  Australian Atomic Energy Commission
        Research Establishment, Lucas Heights
        Sutherland 2232, N.S.W. *Australia*

(1982)  Shanghai Institute of Metrology Technique
        Chang Le Road, Shanghai
        *People's Republic of China*

(1976)  Division of Radiological Protection
        Bhabha Atomic Research Centre, Trombay
        Bombay, *India*

(1980)  National Institute of Health
        Seoul 122, *Republic of Korea*

(1979)  Tun Ismail Atomic Research Centre, PUSPATI
        Bandar Baru Bangi
        Bangi, Selangor, *Malaysia*

(1978)  Pakistan Institute of Nuclear Science and Technology
        Post Office Nilare, Rawalpindi
        *Pakistan*

(1976)  Radiation Health Office of the Ministry of Health
        Rizal Avenue, Manila, *Philippines*

(1970)  Radiotherapy Department, General Hospital
        *Singapore*

(1973)  Division of Radiation Protection Service
        Department of Medical Sciences
        Bangkok, *Thailand*

(1977)  Office of Atomic Energy for Peace
        Bangkhen, Bangkok
        *Thailand*

## 5. HYDROLOGY AND INDUSTRY

### 5.1 Isotope hydrology

The application of radioisotopes and radiation in the study of soil moisture and water management has been described in section 3.2. Since the availability of radio-isotopes, the nuclear technique to hydrological problems had been applied initially to the measurement of flow rates of rivers. Though non-nuclear methods are available, the nuclear techniques are particularly suited for fast-flowing turbulent rivers and have the advantage of following sediment movement. Radioisotopes had been used as tracers for studying seepage through dams, measurement of flow rate and direction of ground-water flow.

The study of isotope hydrology can be divided into two main branches — the environmental isotope hydrology and the artificial isotope hydrology. The former deals with natural isotope variations established in water. The natural isotopes measurement involves analyses for radioactive tritium and carbon-14 and stable isotopes, deuterium and oxygen-18. Isotope hydrology methods permit determination of origin of ground water and assessment of the recharge of the reservoir. In the development of the isotope hydrology techniques, IAEA collaborated with many international organizations and those in the United Nations system. The IAEA isotope hydrology laboratory in Vienna has become the focal point for world-wide hydrological data analysis.

In the studies of water resources, suitable artificial radioisotopes are used in injection into the water system under investigation. The total volume of underground water and the direction of water flow can be studied. Nuclear sedimentology is the application of nuclear techniques to the study of erosion and build-up of sediments. A great deal of effort has been expended in the past years in developing Cs-137 and Pb-210 isotope-techniques for studying erosion and sedimentation. The primary aim of these studies is to measure sedimentation rates as a function of time and to correlate any changes with historical events such as land-use change, road building or mining. Then a rational basis for a programme of soil conservation could be established.

In many developing countries in the ESCAP region, there are serious water supply problems, particularly in the cities. For example: water tables are falling; artesian bores have lost pressure; wells are running dry; surface-water storage areas, subject to monsoons, have been filled in because of buildings and road construction; urban development has reduced the ability of rainwater to percolate into the ground, leading to more run-off; burgeoning populations and industrial development demand even more water; rivers and waterways suffer more from industrial pollution; natural and underground water storages (aquifers) may be contaminated by polluted recharge water or may deteriorate owing to sea-water ingress resulting from over-exploitation; changes in agricultural practices enhance soil erosion and lead to accelerated sedimentation of lakes and reservoirs. In short, more people are demanding more water from a smaller resource which may be deteriorating in quality. Government agencies in a number of countries are tackling these problems, but, for many countries in the South-East Asian region, an important limitation has been a lack of systematic measurements. A water-management policy can only be developed and put into practice when enough measured data on the various water resources have been obtained.

IAEA has assisted and given guidance on isotope hydrology techniques in many developing countries. For example, stable isotope and tritium analyses were applied in the studies on the volcanic island of Cheju, Republic of Korea, to characterize ground waters with respect to time and place of recharge and to determine the nature of mixing of the different ground-water sources and to estimate their residence times. Similar hydrological studies have been carried out in Pakistan. IAEA, the French Commissariat à l'energie atomique and Singapore have been co-operating in the study of the movement of sediments associated with the land reclamation project in Singapore. India has set up facilities for measurement of environmental stable isotopes and radioisotopes with assistance from UNDP and samples have been analysed using the equipment provided.

Since 1978, Australia provided funds for a project on isotope hydrology and sedimentology as part of an IAEA Regional Co-operative Agreement for countries in the South-East Asian region. Members of this agreement were Australia, Bangladesh, Indonesia, Malaysia, Philippines, Republic of Korea, Sri Lanka and Thailand. The project emphasized the application of environmental isotope techniques in the study of ground-water problems and research on the use of Cs-137 in studies of soil erosion and sedimentation. Environmental tritium analytical facilities were set up in Indonesia and in the Republic of Korea and C-14 analytical facilities were made operational in Indonesia. Scientists from these countries had been trained at a research establishment of the Australian Atomic Energy Commission. One co-ordinated research programme was set up to study ground water in and around Bangkok, Jakarta and Seoul. Some studies had also been carried out in Malaysia and Sri Lanka. Although the Australian financial support ceased in June 1983, a number of co-operative isotope hydrology and sedimentology projects are receiving continued support under the IAEA regular programme of technical co-operation.

## 5.2 Industrial isotope-radiation technology

Radioisotopes and radiation technologies are increasingly being applied in industry. Since they are capable of saving energy and resources, the applications of the technology are important for industrial modernization. During the past decade development of new applications and improvement in the reliability of instruments and equipment have brought about an increase in the use of the isotope-radiation technology in industry.

Potential applications in the chemical and food industries are the uses of radiation technology for immobilization of biologically active materials such as enzymes, antibodies, antigens, and tissue cells. The advantage of such treatment is that bio-components are not deactivated and are not contaminated since immobilization can be carried out at low temperatures.

The neutron technique along with natural gamma-ray spectra logging have been widely used in the exploration of oil and gas as well as for uranium deposits. The current trend is to apply the neutron technique for on-line measurement in coal mining.

In sewage treatment, the sewage sludge from wastewater treatment plants can be disinfected by irradiating it using cobalt-60 source and then recycled as fertilizer to recondition soil for crops growing.

On industrial radiation processing, the industrial use of radiation technology is steadily growing in various manufacturing industries such as the production of heat-resistant wires, heat-shrinkable films for food packaging, electrical insulation, vulcanization of tyre rubber, etc.

On sterilization of medical supplies, examples of those products which may now be sterilized by radiation treatment using cobalt-60 include absorbent cotton wool, rubber and plastic sheets, examination and surgical gloves, absorbable surgical sutures (catgut), non-absorbable surgical sutures (silk, nylon, polystyrene, metallic wires), medical tubings, disposable syringes, etc. In a radiosterilization facility, hermetically sealed supplies, packed in cardboard boxes, are made to pass several times before a cobalt-60 source via a conveyor system. Radiation emitted by the cobalt-60 penetrates the packing material, destroying microbial pathogens in the product thus rendering it sterile.

The technique provides a high degree of sterility assurance, no toxic chemicals are used, and it provides capability in many single-use items made of plastic materials.

In industry radioisotopes have been used as tracers to investigate material transport in processing systems in which the flow and mixing patterns are of keen interest in both design and operation.

Nuclear instruments for process control have been much developed. In plant operations, on-line measurements of process parameters or control of product specifications can be carried out with nucleonic gauges. Gauging operations are used in the level detection of liquids or solids in storage, density readings of fluid and slurries, thickness measurement of sheet materials and determination of moisture content. The scope ranges from civil construction as in controlling the density and moisture in concrete and soils to the control of thickness in the high-speed production of paper, plastic films and metal sheets in fully automated plants. The thickness can be measured and precisely controlled without any physical contact with the materials whether they may be hot, damp, soft or plastic.

The nucleonic gauging system improves product uniformity, saves raw materials and reduces off-grade production. It was reported that the cost-to-benefit ratios in installation of nucleonic gauges were as follows:[11]

| Instruments | Cost-to-benefit ratio |
|---|---|
| Plastic film thickness gauge | 1 : 3 |
| Paper thickness and moisture content gauge | 1 : 9 |
| Zinc-coating gauge | 1 : 30 |
| Sulphur-in-oil gauge at desulphurization plant | 1 : 10 |
| Moisture-in-coke gauge for blast furnace | 1 : 20 |
| Leak detection of domestic central heating-system by radiotracer | 1 : 7 |

The improvement on reliability and safety of machines and systems would prevent serious accidents or breakdown in production lines. Isotope radiography based on irridium-192, caesium-137 and cobalt-60 has been used extensively in the quality control of weldings and castings for machinery,

---

[11]  *The IAEA Bulletin*, vol. 23, No. 4 (December 1981).

pipelines and boilers. The importance of non-destructive testing (NDT) by radiography has been demonstrated in big construction projects for nuclear and fossil-fuel power plants, petroleum refining, petrochemical plants and cross-country pipelines transporting natural gas and other petroleum products.

IAEA has successfully promoted radiosterilization with the assistance of UNDP in India and the Republic of Korea. The cobalt-60 loading of the ISOMED plant in India has been upgraded to 250 kcuries and the plant continues to offer radiation services to hospitals, manufacturers of medical products and radiopharmaceutical industries. Varieties of medical products have been sterilized at the plant. The same applies to the plant in the Republic of Korea.

A large demonstration project on radiation technology is being implemented in Indonesia by IAEA under the often-mentioned RCA/UNDP project. The transfer of technology to participating RCA members, developing countries in the ESCAP region, involves radiation vulcanization of natural rubber, electron-beam curing of surface coating on wood products and cross-linking of wire insulation. A semi-commercial scale demonstration plant, 300

kCi, for rubber vulcanization using cobalt-60 and for surface coating using electron-beam accelerator has been commissioned in Indonesia since 1983. The plants which could be also used for sterilization of medical supplies will provide on-the-job training and market development in the RCA member States and the emphasis is on establishing commercial irradiation services and on demonstrating economic viability.

IAEA is also executing a sub-project of the so-called RCA/UNDP project on the nucleonic control systems used in the steel industry in India and in the paper industry at the Siam Kraft Paper Company in Thailand. In the latter case the first year of operation of the demonstration plant showed that the nucleonic control system can pay for itself in less than one year mainly as a result of resource and energy savings.

On non-destructive testing, IAEA has provided technical assistance to Indonesia, Malaysia, Pakistan, the Philippines, Sri Lanka and Thailand by providing equipment, experts and fellowships in order to develop the necessary skills on the technique of radiography. With the assistance of UNDP in the RCA programme, three extensive training courses on radiographic and ultrasonic inspections are being conducted in the region until 1987.

**Part Three**

**CO-OPERATION IN NUCLEAR SCIENCE
AND TECHNOLOGY**

# 6. INTER-REGIONAL AND REGIONAL CO-OPERATION

## 6.1 Interregional co-operation

The development of science in less developed and in most of the developing countries in the ESCAP region has been an irregular and variable process. It depends on the national budget and funds available for special areas of science and on many other factors. As a result developing countries do not have the broad capability in scientific disciplines which one expects in developed countries, but they have potential in certain fields where they could eventually reach international standards. The fields of peaceful uses of nuclear energy are highly diversified. At present programmes for the use of nuclear science and technology in countries in the ESCAP region vary from virtually non-existent to highly developed stages.

Combining limited potentials of various developing countries in the region may lead to overall scientific strength and can contribute to an accelerated development in science, lead to self-reliance and eventually reduce external support from developed countries. Developing countries in the ESCAP region may have common interests owing to similarities in climate, geography and economic structures. Projects of practical benefit to most of them may be defined for joint scientific efforts. Based on mutual acceptance, exchange of information and scientists, who are established in the region, would reduce travel costs to distant developed countries. Already acquired expensive equipment and expertise attained in certain institutions would be better utilized on a regional basis through regional co-operation.

In its efforts to promote transfer of nuclear science and technology, the International Atomic Energy Agency (IAEA) awarded research contracts to institutes in particular in developing countries. Later many research contracts were organized in the form of co-ordinated research projects. Such co-ordinated research projects provide opportunities for direct contact among scientists working in the same field in the region as well as with other regions as the case may be. The research subjects are usually oriented towards the needs of the developing countries and the preference in making awards is given to institutes located in these countries. IAEA acts upon applications received directly from research institutes. By the end of the past decade about 120 co-ordinated research projects had been completed. Some three quarters of them were in the field of radioisotopes and radiation application. For all regions, the current award rate for new contracts is of the order of 150 a year and research co-ordination meetings are held each year.

With the funds available, and with those from the United Nations Development Programme and international organizations and developed countries, IAEA executed interregional and regional training courses in the field of nuclear science and technology at various host institutions. Scientists from different parts of the world were selected to attend such courses. Specialists from developed countries and qualified scientists from developing countries were invited to give lectures, to lead seminars and to conduct practical classes. In this way, scientists from different countries gain contact and co-operate with each other and with scientists of the developed countries who work in similar fields. Such arrangements bring about desirable international co-operation. The interregional and regional training courses during 1983 are listed in table 6.1.

### Table 6.1. Regional and interregional activities

| Year | Project title and code | Place(s) and dates | Source of funds |
|------|------------------------|--------------------|-----------------|
| 1983 | Regional training course on nuclear instrumentation RAP/4/002 | Lusaka, Zambia 28 February to 21 April | Agency |
| | Second UNDP (RCA) training/demonstration workshop on the use of nucleonic control systems in the paper industry RAS/8/018 | Ban Pong, Thailand, and Tokyo, Japan 28 March to 16 April | UNDP Japan |
| | Interregional training course on radiological protection and nuclear safety INT/9/046 | Buenoe Aires, Argentina 4 April to 30 November | Agency |

**Table 6.1.** *(continued)*

| Year | Project title and code | Place(s) and dates | Source of funds |
|------|------------------------|--------------------|-----------------|
| | Interregional training course on the induction and use of mutations in plant breeding INT/5/089 | Seibersdorf, Austria 6 April to 19 May | Agency |
| | Interregional training course on uranium ore processing INT/3/012 | Madrid, Spain 11 April to 13 May | Agency |
| | Interregional training course on quality assurance INT/4/063 | Saclay, France 11 April to 18 May | Agency |
| | Interregional training course on electric system expansion planning INT/0/029 | Argonne, Illinois, United States 18 April to 17 June | Agency |
| | Study tour on nuclear power development INT/0/033 | Bulgaria, Czechoslovakia, USSR, German Democratic Republic and Hungary 7 May to 4 June | Agency |
| | Interregional training course on the use of isotope and radiation techniques in studies on soil/plant relationships INT/5/086 | Seibersdorf, Austria 16 May to 1 July | Agency SIDA |
| | Interregional training course and study tour on induced mutations in plant breeding with special attention to cross-pollinating plant species INT/5/090 | Sofia, Bulgaria, Ukrainian SSR and Byelorussian SSR 16 May to 4 July | Agency |
| | FAO/IAEA interregional training course on the use of nitrogen-15 in soil science and plant nutrition INT/5/085 | Leipzig, German Democratic Republic 25 May to 17 June | Agency |
| | Interregional training course on energy planning in developing countries with special attention to nuclear energy INT/0/032 | Ljubljana, Yugoslavia, and technical visit to Munich, Federal Republic of Germany 6 June to 1 July | Agency |
| | Training workshop on monsoon rainfall prediction INT/1/024 | Dhaka, Bangladesh 20 to 23 June | UNFSSTD |
| | Interregional advanced training course on nuclear electronics INT/4/064 | Berlin (West) 18 July to 14 October | Agency Federal Republic of Germany |
| | Interregional training course and study tour on nuclear techniques in the study of parasitic infections of man INT/6/028 | Bethesda, Maryland, and Atlanta, Georgia, United States 8 to 30 August | Agency |
| | UNDP regional (RCA) industrial training/demonstration workshop on on-stream analysis of mineral concentrators employing nucleonic control systems RAS/8/022 | Lucas Heights, Australia and Quezon City and Baguio City, Philippines 29 August to 17 December | Australia |
| | Interregional training course and study tour on nuclear medicine INT/6/027 | Moscow, USSR 1 September to 31 October | Agency |

**Table 6.1.** *(continued)*

| Year | Project title and code | Place(s) and dates | Source of funds |
|---|---|---|---|
| | Interregional training course and study tour on neutron physics and nuclear data measurements with accelerators and research reactors INT/1/023 | Tashkent, USSR 4 to 30 September | Agency |
| | Interregional training course on waste management in nuclear facilities INT/9/044 | Karlsruhe, Federal Republic of Germany 5 to 30 September | Agency |
| | Interregional training course on uranium prospection INT/3/011 | Skofja Loka, Yugoslavia 5 September to 7 October | Agency |
| | Interregional training course on probabilistic risk methods applied to safety analysis for nuclear power plants INT/9/043 | Argonne, Illinois, United States 6 September to 7 October | Agency |
| | International training course on physical protection of nuclear facilities and materials INT/0/034 | Albuquerque, New Mexico, USA 7 to 29 September | United States of America |
| | First advanced non-destructive testing training course for ultrasonic and radiographic inspection RAS/8/019 | Tokyo, Japan 12 September to 8 October | UNDP Japan |
| | Interregional training course on environmental isotopes in hydrology INT/8/023 | Vienna, Austria 19 September to 7 October | Agency |
| | UNDP regional (RCA) industrial training/demonstration workshop on radiation sterilization of medical products RAS/8/020 | Bombay, India, and Seoul, Republic of Korea 26 September to 14 October | UNDP |
| | UNDP regional (RCA) industrial training/demonstration workshop on radiation vulcanization of natural rubber latex RAS/8/025 | Jakarta, Indonesia 1 October 1983 to 31 March 1984 | UNDP |
| | UNDP regional (RCA) industrial training/demonstration workshop on nucleonic control systems for steel manufacture RAS/8/021 | Bokaro, India and Tokyo, Japan 10 to 29 October | UNDP Japan |
| | Interregional training course on nuclear power planning and feasibility studies INT/0/031 | Saclay, France 10 October to 18 November | Agency |
| | Interregional training course on control and instrumentation of nuclear power plants INT/4/062 | Karlsruhe, Federal Republic of Germany 10 October to 18 November | Agency |
| | Advanced interregional training course on research reactor utilization INT/4/065 | Budapest, Hungary 10 October to 18 November | Agency |
| | Interregional training course on nuclear project management tools and methods INT/0/030 | Argonne, Illinois, United States of America 11 October to 23 November | Agency |

**Table  6.1.**  *(continued)*

| Year | Project title and code | Place(s) and dates | Source of funds |
|------|------------------------|--------------------|-----------------|
| | Interregional training course on the use of nuclear techniques in animal parasitology INT/5/088 | Nairobi, Kenya 17 October to 11 November | Agency |
| | UNDP regional (RCA) on-the-job training workshop on the maintenance of nuclear instruments RAS/8/024 | Tokyo, Japan 1 November to 20 December | UNDP Japan |
| | Regional training course on nuclear techniques in pesticide research RAS/5/014 | Bangkok, Thailand 7 November to 2 December | Agency |
| | Regional training course on the production and control of radiopharmaceuticals RIA/2/002 | Montevideo, Uruguay 21 November to 16 December | Agency |

*Source:*  IAEA

**Table 6.2. IAEA research and technical contracts, awarded or renewed in 1981 to 1983**
(Countries in the ESCAP region)
(In US dollars)

*Source:* International Atomic Energy Agency

| COUNTRY | YEAR | New contracts | Renewals | Agriculture | Food irradiation | Medicine | Dosimetry | Radiation biology | Health-related environmental research | Physics | Industry/chemistry | Hydrology | Nuclear data | Monaco Laboratory | Nuclear power | Radiological safety | Risk assessment | Nuclear materials | Waste management | TOTAL |
|---|---|---|---|---|---|---|---|---|---|---|---|---|---|---|---|---|---|---|---|---|
| Australia | 1981 | 1 | 1 | — | — | — | — | — | — | — | 6 000 | — | — | — | — | — | — | — | — | 6 000 |
| | 1982 | 1 | 2 | — | — | — | — | 700 | — | — | 12 000 | — | — | — | — | — | — | — | — | 12 700 |
| | 1983 | — | 1 | — | — | — | — | — | — | — | 6 600 | — | — | — | — | — | — | — | — | 6 600 |
| Bangladesh | 1981 | 4 | 6 | 18 300 | — | 3 000 | — | — | 3 000 | — | — | — | — | — | — | — | — | — | — | 35 300 |
| | 1982 | 3 | 10 | 29 800 | 4 000 | 9 000 | — | — | 3 000 | — | — | — | 4 000 | — | — | 7 000 | — | — | — | 49 500 |
| | 1983 | 3 | 8 | 17 000 | 8 000 | 11 000 | — | — | — | — | — | — | — | — | — | 3 700 | — | — | — | 48 400 |
| Burma | 1981 | — | — | — | — | — | — | — | — | — | — | — | — | — | — | — | — | — | — | — |
| | 1982 | — | — | — | — | — | — | 3 515 | — | — | — | — | — | — | — | — | — | — | — | 3 515 |
| | 1983 | 1 | — | — | — | — | — | — | — | — | — | — | 4 000 | — | — | — | — | — | — | 4 000 |
| India | 1981 | 8 | 20 | 32 700 | — | 28 700 | 5 000 | 13 500 | 5 000 | 7 710 | — | — | — | — | 8 300 | 16 000 | — | — | 11 000 | 127 910 |
| | 1982 | 13 | 17 | 23 500 | — | 41 600 | 20 900 | 19 885 | 300 | 11 650 | — | 5 000 | 6 500 | 5 600 | 4 240 | 16 500 | — | — | 11 000 | 166 673 |
| | 1983 | 6 | 12 | 18 962 | — | 17 456 | — | 18 900 | 5 000 | 4 000 | 3 000 | — | 2 000 | — | — | 12 101 | — | — | 5 000 | 86 419 |
| Indonesia | 1981 | 1 | 7 | 20 500 | — | — | — | — | 4 000 | — | — | — | — | — | — | — | — | — | — | 24 500 |
| | 1982 | 3 | 9 | 16 000 | 8 000 | 5 000 | 3 500 | — | 5 000 | — | — | 6 500 | — | — | — | — | — | — | — | 44 000 |
| | 1983 | 5 | 6 | 25 000 | 8 000 | 4 500 | — | 4 000 | — | — | 3 000 | — | — | — | — | — | — | — | — | 44 500 |
| Iran (Islamic Republic of) | 1981 | 1 | 2 | — | — | 7 000 | — | 4 000 | — | 4 980 | — | — | — | — | — | — | — | — | — | 15 980 |
| | 1982 | 2 | 2 | — | — | — | — | — | 10 000 | 4 500 | 4 000 | — | — | — | — | — | — | — | — | 18 500 |
| | 1983 | — | 1 | — | — | — | — | — | — | — | 4 500 | — | — | — | — | — | — | — | — | 4 500 |
| Republic of Korea | 1981 | 1 | 7 | 14 500 | — | 4 500 | — | — | — | — | 5 000 | — | — | — | — | — | — | — | — | 30 000 |
| | 1982 | 2 | 5 | 7 000 | — | 4 500 | — | — | — | — | — | 7 500 | — | 3 000 | — | 3 000 | — | 4 000 | 3 500 | 32 500 |
| | 1983 | 5 | 5 | 18 500 | — | 15 100 | — | — | — | — | — | — | — | — | — | 6 500 | — | — | — | 40 100 |
| Malaysia | 1981 | 1 | 3 | 14 000 | — | 4 500 | — | — | — | — | — | — | — | — | — | — | — | — | — | 18 500 |
| | 1982 | 3 | 8 | 29 000 | — | 5 000 | — | 3 000 | 4 000 | — | — | — | — | — | — | — | — | — | — | 41 000 |
| | 1983 | 4 | 5 | 24 500 | — | 4 500 | — | 8 500 | — | — | 3 000 | — | — | — | — | — | — | — | — | 40 500 |
| New Zealand | 1981 | — | — | — | — | — | — | — | — | — | — | — | — | — | — | — | — | — | — | — |
| | 1982 | — | 1 | — | — | — | — | — | — | 4 000 | — | — | — | — | — | — | — | — | — | 4 000 |
| | 1983 | — | — | — | — | — | — | — | — | — | — | — | — | — | — | — | — | — | — | — |

**Table 6.2.** *(continued)*

| COUNTRY | YEAR | New contracts | Renewals | Agriculture | Food irradiation | Medicine | Dosimetry | Radiation biology | Health-related environmental research | Physics | Industry/chemistry | Hydrology | Neclear data | Monaco Laboratory | Neclear power | Radiological safety | Risk assessment | Nuclear materials | Waste management | TOTAL |
|---|---|---|---|---|---|---|---|---|---|---|---|---|---|---|---|---|---|---|---|---|
| Pakistan | 1981 | 2 | 2 | 6 000 | 5 000 | 6 000 | — | — | — | 4 000 | — | — | — | — | — | — | — | — | — | 17 000 |
| | 1982 | 4 | 3 | 8 000 | 5 000 | 2 000 | — | 5 000 | — | — | — | — | 3 000 | — | — | 3 500 | — | — | — | 27 000 |
| | 1983 | 7 | 7 | 27 166 | — | 11 000 | — | 3 000 | — | — | — | — | 3 000 | — | — | 7 800 | — | — | — | 51 166 |
| Philippines | 1981 | 1 | 6 | 4 870 | 3 000 | 10 800 | — | 3 500 | — | — | — | — | — | — | — | — | — | — | — | 29 970 |
| | 1982 | 4 | 4 | 6 500 | 4 000 | 8 300 | — | 2 500 | — | — | — | 3 500 | — | — | — | — | — | — | — | 26 300 |
| | 1983 | 9 | 4 | 20 500 | 7 000 | 11 700 | — | — | — | — | — | — | — | — | — | 4 000 | 6 000 | — | — | 54 200 |
| Singapore | 1981 | — | — | — | — | — | — | — | — | — | — | — | — | — | — | — | — | — | — | — |
| | 1982 | 1 | 1 | — | 3 000 | — | — | — | — | — | — | — | — | 5 000 | — | — | — | — | — | 3 000 |
| | 1983 | 1 | 1 | — | 3 000 | — | — | 4 000 | — | — | — | — | — | 5 000 | — | — | — | — | — | 7 000 |
| Sri Lanka | 1981 | 3 | 5 | 19 000 | — | 17 800 | — | — | — | — | — | — | — | — | — | — | — | — | — | 36 800 |
| | 1982 | 2 | 5 | 24 240 | — | 4 500 | — | 3 000 | — | — | — | — | — | — | — | — | — | — | — | 31 740 |
| | 1983 | 5 | 5 | 31 000 | — | 4 500 | — | 8 400 | — | — | — | — | — | — | — | — | — | — | — | 43 900 |
| Thailand | 1981 | 4 | 8 | 17 500 | 12 000 | 5 000 | — | 7 500 | — | — | — | — | — | — | — | — | — | — | — | 45 000 |
| | 1982 | 10 | 9 | 21 500 | 9 000 | 11 500 | 3 000 | 7 000 | — | — | — | 12 000 | 5 000 | — | — | — | — | — | — | 71 110 |
| | 1983 | 14 | 14 | 43 500 | 8 000 | 29 200 | 5 110 | 16 000 | — | — | 3 000 | 6 000 | 5 000 | — | — | — | 5 000 | — | — | 115 700 |
| Viet Nam | 1981 | — | — | — | — | — | — | — | — | — | — | — | — | — | — | — | — | — | — | — |
| | 1982 | — | — | — | — | — | — | — | — | — | — | — | — | — | — | — | — | — | — | — |
| | 1983 | 2 | — | — | — | 5 000 | — | — | — | — | — | — | 4 000 | — | — | — | — | — | — | 9 000 |

Table 6.2 shows the distribution of funds, by subject matter and countries, in the ESCAP region, for the IAEA research and technical contracts, awarded and renewed in 1981 to 1983.

It could be taken also that the award of co-ordinated research contracts results in countries' requests for technical assistance (later, technical co-operation) programmes from IAEA for further support in terms of equipment, supplies, expert advice and fellowships training. Depending on individual countries, they resulted in the formulation of larger-scale projects financed by UNDP or international organizations. The UNDP projects in the ESCAP region for 1980-1984, executed by IAEA, are as follows:

| Recipient country and title of project | Start of operation | Project duration (Years) | UNDP contribution (In US dollars) |
|---|---|---|---|
| Philippines | | | |
|     Philippine nuclear manpower development programme | September 1981 | 3.5 | 1 019 850 |
| Region of Asia and the Far East | | | |
|     Support for regional co-operation in the industrial application of isotopes and radiation technology | May 1982 | 8.0 | 4 381 516 |
| Indonesia | | | |
|     Application of isotopes and radiation to increasing agricultural production | July 1982 | 5.0 | 1 560 000 |
| Iran (Islamic Republic of) | | | |
|     Pilot demonstration plant for radiosterilization and other application of radiation technology | December 1982 | 3.5 | 1 604 000 |

The International Atomic Energy Agency's distribution of technical co-operation inputs by field and region, 1980-1983 (in thousands of US dollars) could be summarized as follows:

| Field of activity | Africa | Asia and the Pacific | Europe | Latin America | Middle East | Inter-regional | All regions |
|---|---|---|---|---|---|---|---|
| General atomic energy development | 2 349.9 | 1 046.6 | 549.6 | 2 929.1 | 109.8 | 1 088.3 | 8 073.3 |
| Nuclear physics | 1 280.0 | 2 671.3 | 1 266.1 | 1 955.0 | 309.4 | 1 050.6 | 8 532.4 |
| Nuclear chemistry | 506.8 | 1 233.5 | 950.8 | 458.6 | 42.2 | 216.0 | 3 407.9 |
| Prospecting, mining and processing of nuclear materials | 2 722.8 | 1 486.4 | 885.5 | 3 538.7 | 3.5 | 328.7 | 8 965.6 |
| Nuclear engineering and technology | 779.3 | 3 002.5 | 4 273.1 | 4 234.3 | 298.6 | 1 511.0 | 14 098.8 |
| Agriculture | 6 312.5 | 4 695.9 | 1 121.6 | 3 923.7 | 55.2 | 1 660.3 | 17 769.2 |
| Medicine | 1 914.8 | 2 316.6 | 510.9 | 2 528.9 | 143.3 | 1 233.0 | 8 647.5 |
| Biology | 197.7 | 221.6 | 492.3 | 84.8 | 7.4 | 107.4 | 1 111.2 |
| Industry and hydrology | 1 689.3 | 5 289.6 | 1 113.1 | 1 258.2 | 73.9 | 69.6 | 9 493.7 |
| Safety in nuclear energy | 2 228.5 | 2 365.3 | 1 754.6 | 1 499.5 | 107.2 | 1 277.3 | 9 232.4 |
| TOTAL | 19 981.6 | 24 329.3 | 12 917.6 | 22 410.8 | 1 150.5 | 8 542.2 | 89 332.0 |

**Table 6.3. Financial summary: 1958-1983 on assistance provided to International Atomic Energy Agency (IAEA) member States in the ESCAP region**
(In thousands of US dollars)

| Recipient | Assistance provided, by type | | | | Assistance provide, by source | | | | |
|---|---|---|---|---|---|---|---|---|---|
| | Experts | Equipment | Follow-ships | TOTAL | UNDP | IAEA funds | Extra budgetary funds [a] | In kind | TOTAL |
| Afghanistan | 374.1 | 386.8 | 120.5 | 881.4 | 92.9 | 706.7 | – | 81.8 | 881.4 |
| Bangladesh | 718.0 | 1 442.2 | 1 435.9 | 3 596.1 | 63.0 | 1 475.6 | 969.5 | 1 088.0 | 3 596.1 |
| Burma | 746.4 | 944.1 | 200.6 | 1 891.1 | 537.0 | 1 250.5 | – | 103.6 | 1 891.1 |
| Hong Kong | 59.9 | 105.2 | 26.1 | 191.2 | – | 182.2 | – | 9.0 | 191.2 |
| India | 997.9 | 3 657.8 | 2 541.7 | 7 197.4 | 2 920.3 | 1 280.7 | 1 834.2 | 1 162.2 | 7 197.4 |
| Indonesia | 1 417.4 | 1 374.7 | 1 049.5 | 3 841.6 | 935.7 | 1 882.9 | 333.9 | 689.1 | 3 841.6 |
| Iran (Islamic Republic of) | 679.9 | 747.5 | 482.5 | 1 909.9 | 1 129.1 | 497.9 | 9.5 | 273.4 | 1 909.9 |
| Malaysia | 655.5 | 1 017.5 | 620.8 | 2 293.8 | 1.6 | 1 345.7 | 488.0 | 458.5 | 2 293.8 |
| Mongolia | 141.5 | 685.9 | 17.2 | 844.6 | – | 827.4 | 10.6 | 6.6 | 844.6 |
| Niue | 7.8 | 6.9 | – | 14.7 | 14.7 | – | – | – | 14.7 |
| Pakistan | 1 463.2 | 2 167.3 | 2 378.4 | 6 008.9 | 1 842.0 | 2 839.0 | 90.5 | 1 237.4 | 6 008.9 |
| Philippines | 1 261.4 | 2 126.4 | 2 410.0 | 5 797.8 | 1 383.4 | 2 132.8 | 610.0 | 1 671.6 | 5 797.8 |
| Republic of Korea | 1 504.4 | 1 219.2 | 1 816.0 | 4 539.6 | 566.8 | 2 024.0 | 589.6 | 1 359.2 | 4 539.6 |
| Singapore | 193.0 | 620.0 | 71.3 | 884.3 | – | 730.4 | 101.1 | 52.8 | 884.3 |
| Sri Lanka | 668.1 | 1 266.6 | 965.9 | 2 900.6 | 305.4 | 1 868.4 | 321.0 | 405.8 | 2 900.6 |
| Thailand | 1 216.7 | 1 542.4 | 2 398.2 | 5 157.3 | 545.5 | 2 354.1 | 654.6 | 1 603.1 | 5 157.3 |
| Viet Nam | 183.9 | 1 550.8 | 401.8 | 2 136.5 | 31.4 | 1 795.8 | 18.1 | 291.2 | 2 136.5 |
| *Interregional projects and training courses* | | | | | | | | | |
| Asia and the Pacific | 1 673.7 | 1 829.1 | 862.8 | 4 365.6 | 2 796.5 | 744.6 | 359.0 | 465.5 | 4 365.6 |
| Interregional | 3 879.7 | 2 205.6 | 10 770.7 | 16 856.0 | 1 790.5 | 11 864.8 | 1 588.4 | 1 612.3 | 16 856.0 |

*Source:* IAEA Technical Assistance Report 1983.

Table 6.3 shows the financial summary (1958-1983) on assistance to IAEA member States in the ESCAP region.

## 6.2 Regional co-operation in Asia and the Pacific

Under the auspices of IAEA, a training and research programme for the development of nuclear science through regional collaboration in Asia and the Pacific was established in 1964 by an agreement between India, Philippines and the IAEA. India provided a neutron crystal spectrometer which was installed at the Philippines research reactor centre. Apart from India and the Philippines, Indonesia, the Republic of Korea and Thailand co-operated in the five-year programme for training and research in solid state physics using the neutron crystal spectrometer.

A Regional Co-operative Agreement for Research, Development and Training Related to Nuclear Science and Technology (RCA) was formulated during 1970-1971 and entered into force on 12 June 1972 for a five-year period and 11 Governments were parties to it : Bangladesh, India, Indonesia, Malaysia, Pakistan, Philippines, Republic of Korea, Singapore, Sri Lanka, Thailand and Viet Nam.

RCA was further extended in June 1977 and is currently in force between IAEA and the 11 original parties and in addition, Australia and Japan.

The purpose of the Agreement is to promote and coordinate research, development and training projects in nuclear science and technology through co-operation between appropriate national institutions. Any party may initiate a co-operative project by means of a written proposal to IAEA. If at least two parties are interested in participating in such a project, the interested Governments and IAEA will initiate negotiations with a view to establishing the project. Given the consent of the parties to an established project, any other State may also participate in the project. The progress and the co-ordination of the programmes are reviewed at annual meetings of parties to the Agreement. Proposals for establishing new co-operative projects are also considered at such meetings.

IAEA support for such RCA projects have been limited to its research contract funds and its covering of certain costs of meetings. With the participation of Australia and Japan and recently also of India the programme has been assisted greatly by funds, equipment, facilities and expert services made available by these countries.

The IAEA-RCA co-operative research projects include the following (though some of them have been completed and in some of them all parties to RCA take part):

(a) The use of ionizing radiation for the preservation of fish and fishery products began in 1975. During 1981-1983 the project had financial contributions from Japan and became known for short as the Food Irradiation project.

(b) The use of radiation-induced mutation for the improvement of grain legume production.

(c) Sterilization of biological tissue grafts.

(d) Neutron scattering studies.

(e) Health-related environmental research using nuclear techniques.

(f) Isotope hydrology and sedimentology. Australia provided funds for the project.

(g) Semi-dwarf mutants for rice improvement.

(h) Basic science using research reactors. India provided contributions to a three-week workshop on the use of micro-processors in research reactor utilization held in India in 1984.

(i) Nuclear instrument maintenance. Later this became a sub-project in the UNDP industrial project.

In 1983 IAEA set up the following co-ordinated research projects:

(a) Improvement of cancer therapy;

(b) Nuclear medicine for thyroid and liver diseases;

(c) Nuclear techniques for tropical parasitic diseases;

(d) Development of technitium-99m generator systems.

The first two projects have financial support from Japan. The Japanese Government announced an offer to IAEA of a teletherapy apparatus for uterine cancer treatment to be given to one of the RCA countries to help create a regional centre of excellence in this field.

The financial allocations for the above-mentioned RCA projects are tabulated in table 6.4 (ref. [1]). Countries participating in various RCA projects are tabulated in Table 6.5.[1]

In 1977, India, Indonesia, Malaysia, Pakistan, Philippines, Republic of Korea, Singapore, Sri Lanka and Thailand separately requested IAEA technical assistance on industrial isotopes and radiation applications. In 1978, for the first time IAEA launched a regional (Asia and the Pacific) technical assistance project to give an assessment on the requirements of Governments and industries in Asia and the Pacific in the use of radioisotopes and radiation techniques to bring about economic and social benefits. As the outcome of the studies, the above-mentioned Governments requested further assistance from the UNDP regional project entitled "Support for regional co-operation in the industrial application of isotopes and radiation technology" which started in May 1982 though the preparatory assistance started in August 1979. This project has sometimes been referred to as the RCA/UNDP project. The Governments participating in this project include Australia, Bangladesh, India, Indonesia, Japan, Malaysia, Pakistan, Philippines, Republic of Korea, Singapore, Sri Lanka and Thailand. Australia and Japan participate by giving assistance in cash and in kind. The contributions for the project are from UNDP, US$ 4,381,516; Governments, US$ 6,427,457; and industries US$ 1,653,440. IAEA is the executing agency of this project which will last until 1987.

The project involves technology transfer in five selected fields: industrial tracer applications; non-destructive testing; nucleonics control systems; radiation processing and nuclear instruments maintenance.

The financial allocations per year are tabulated in table 6.6.[2]

While the above review covers ongoing activities and there can already be a source of satisfaction for the regional countries, the purpose of this volume is also to analyse future needs. For this purpose a group of regional experts was called together during 14-17 January 1985. In the next part of the volume the report of the Expert Group Meeting shall be presented.

---

[1] M. Kobayashi: *IAEA Bulletin*, vol. 26, No. 1 (March 1984), p. 22-27.

[2] M. Kobayashi: *IAEA Bulletin*, vol. 26, No. 1 (March 1984), p. 22-27.

## Table 6.4. Funds allocated for RCA activities 1978-1984
### (In US dollars)

| Title of project/activity | 1978 | 1979 | 1980 | 1981 | 1982 | 1983 | 1984 | TOTAL (by activity) |
|---|---|---|---|---|---|---|---|---|
| Use of induced mutations for the improvement of grain legume production | 19 000 | 49 309 | 49 200 | 81 500 | 71 000 | 80 000 | 73 000 | 423 009 |
| Food irradiation | 8 000 | 27 400 | 76 000 | 80 000 | 80 000 | 40 000 | 82 500 | 393 900 |
| Use of nuclear techniques in improving buffalo production | 28 000 | 50 243 | 50 200 | 70 700 | 52 000 | 44 000 | 85 000 | 380 143 |
| Radiation sterilization of medical supplies | – | 30 000 | 51 000 | 35 000 | 39 000 | 35 000 | 30 000 | 220 000 |
| Health-related environmental research | 18 260 | 6 000 | 20 000 | 44 000 | 48 000 | 30 000 | 74 000 | 240 260 |
| Maintenance of nuclear instruments | | 52 700 | 47 500 | 53 500 | 65 000 | 45 000 | 60 000 | 323 700 |
| Neutron scattering | 27 500 | 35 400 | 23 000 | 12 700 | – | – | – | 98 600 |
| Basic science using research reactors | | | | | | 40 000 | 40 000 | 80 000 |
| Isotope applications in hydrology and sedimentology | | 74 447 | 105 300 | 105 000 | 95 000 | 55 000 | 25 000 | 459 747 |
| Semi-dwarf mutants for rice improvement | – | – | – | – | 50 000 | 68 000 | 73 000 | 191 000 |
| Industrial applications of isotopes and radiation technology (UNDP) | 92 085 | 24 892 | 123 798 | 2 284 753 | 2 996 626 | 2 759 668 | 1 802 759 | 10 084 581 |
| Improvement of cancer therapy | – | – | – | – | – | 48 000 | 130 000 | 178 000 |
| Nuclear medicine for thyroid and liver diseases | – | – | – | – | – | 30 000 | 155 000 | 185 000 |
| Nuclear techniques for tropical parasite diseases | – | – | – | – | – | 31 000 | 40 000 | 71 000 |
| Development of Tc-99m generator systems | – | – | – | – | – | 22 000 | 50 000 | 72 000 |
| RCA Working Group meetings | – | – | – | 3 600 | 4 000 | 4 000 | 4 000 | 15 600 |
| TOTAL | 192 845 | 350 391 | 545 999 | 2 770 753 | 3 500 626 | 3 331 668 | 2 724 259 | 13 416 540 |

## Table 6.5. RCA regional co-operative projects

| Project title | Australia | Bangladesh | India | Indonesia | Japan | Malaysia | Pakistan | Philippines | Republic of Korea | Singapore | Sri Lanka | Thailand | Viet Nam |
|---|---|---|---|---|---|---|---|---|---|---|---|---|---|
| 1. Use of induced mutations for the improvement of grain legume production | | x | x | x | | x | x | x | x | | x | x | |
| 2. Food irradiation | | x | x | x | x | x | x | x | x | | x | x | (x) |
| 3. Use of nuclear techniques in improving buffalo production | x | x | x | x | | x | | | x | x | x | x | |
| 4. Radiation sterilization of medical supplies | x | x | x | x | | | x | x | x | | | x | |
| 5. Health-related environmental research | | x | x | x | x | x | x | x | x | x | | x | |
| 6. Maintenance of nuclear instruments | | x | x | x | | x | x | x | x | | x | x | (x) |
| 7. Isotope applications in hydrology and sedimentology | x | | | x | | x | | | x | | | x | x |
| 8. Semi-dwarf mutants for rice improvement | | x | x | x | x | x | x | x | x | | x | x | x |
| 9. Basic science using research reactors | | x | x | x | | x | | x | x | | x | x | x |
| 10. Industrial applications of isotopes and radiation technology (UNDP) | x | x | x | x | x | x | x | x | x | x | x | x | (x) |
| *11. Cancer therapy | | x | x | | x | x | x | x | x | x | x | x | |
| *12. Nuclear medicine | x | x | x | x | x | x | x | x | x | x | x | x | x |
| *13. Parasitic diseases | | x | x | x | | x | x | x | | | x | x | |
| *14. Tc-99m generators | x | x | x | x | | x | x | x | x | | x | x | |

(subject to negotiation)                                                                 * expected participants.

## Table 6.6 Resources allocated for the industrial applications project as of January 1984
### (In US dollars)

| | 1978 | 1979 | 1980 | 1981 | 1982 | 1983 | 1984 | 1985 | 1986 | 1987 |
|---|---|---|---|---|---|---|---|---|---|---|
| Technical co-operation programme | 92 085 | 11 357 | – | – | – | – | – | – | – | – |
| UNDP funds | – | 13 535 | 123 798 | 1 043 454 | 910 347 | 724 983 | 630 200 | 514 450 | 294 900 | – |
| Countributions: Australia and Japan* | – | – | – | 213 110 | 618 050 | 987 617 | 337 646 | 279 472 | 44 631 | – |
| In kind contributions: | | | | | | | | | | |
| Other participating Governments | – | – | – | 572 689 | 1 053 229 | 899 800 | 678 644 | 454 855 | 278 714 | – |
| Participating Industries | – | – | – | 455 500 | 415 000 | 147 268 | 147 269 | 145 829 | 145 819 | 196 755 |
| TOTAL | 82 085 | 24 892 | 123 798 | 2 284 753 | 2 996 626 | 2 759 658 | 1 802 759 | 1 394 606 | 764 064 | 196 755 |

*  In cash and in kind.

Part Four

# THE FUTURE: REPORT AND RECOMMENDATIONS OF A REGIONAL EXPERT GROUP

# 7. ORGANIZATION OF THE MEETING

The Regional Expert Group Meeting for the United Nations Conference for the Promotion of International Co-operation in the Peaceful Uses of Nuclear Energy was convened at Bangkok from 14 to 17 January 1985.

## Attendance

Fifteen experts from twelve ESCAP member countries attended the Meeting. For the list of participants see the annex to this report.

The secretariat of the United Nations Conference for the Promotion of International Co-operation in the Peaceful Uses of Nuclear Energy was represented at the Meeting. Representatives of the Food and Agriculture Organization of the United Nations (FAO), the International Atomic Energy Agency (IAEA) and the World Health Organization (WHO) were also in attendance.

## Election of officers

The Meeting elected Mr. Mohd. Ghazali, Director General, Nuclear Energy Unit, Kuala Lumpur, Malaysia as Chairman. Mr. K. Sundaram, Director, Bio-Medical Group, Bhabha Atomic Research Centre, Bombay, India, and Mr. Xu Guanren, Director-General, Institute for Application of Atomic Energy, Beijing, China, were elected Vice-Chairmen. Mr. Gary Makasiar, Director, Planning Service, Ministry of Energy, Manila, Philippines, was elected Rapporteur of the Meeting.

## Adoption of the agenda

The Meeting adopted the following agenda:

1) Opening of the Meeting.

2) Election of officers.

3) Adoption of the agenda.

4) Nuclear power.

5) Application of radioisotopes and radiation.

6) Conclusions and recommendations for the Conference.

7) Adoption of the report.

## Opening of the Meeting

The Meeting was opened by the Executive Secretary of ESCAP. In welcoming the participants, the Executive Secretary expressed his special gratitude to Ambassador Amrik S. Mehta, Personal Representative of the Secretary-General of the United Nations and Secretary-General of the United Nations Conference for the Promotion of International Co-operation in the Peaceful Uses of Nuclear Energy.

The Executive Secretary outlined the tasks of the Meeting which were to assess the constraints the region faced and to formulate recommendations for realizing the potential benefits of scientific and technological progress in the nuclear field. The recommendations of the Meeting would constitute a useful regional input to the forthcoming global Conference.

The Executive Secretary stressed that the involvement of ESCAP in the field of nuclear energy was guided by General Assembly resolutions, by direct requests from the Conference secretariat, and by the interest of some of the member countries in developing nuclear energy for electric power. He also referred to IAEA success in fostering interaction among Asian and Pacific countries through regional co-operative agreements. The Executive Secretary drew attention to the fact that ESCAP was the first of the regional commissions to hold a preparatory expert group meeting which thus represented a pioneering effort. He assured the Meeting that the secretariat was ready to assist the Meeting in any possible way.

The Secretary-General of the Conference, Ambassador Amrik S. Mehta, expressed his appreciation to the Executive Secretary, and described the background and genesis of the Conference which was first considered at the thirty-second session of the United Nations General Assembly. He pointed out that the General Assembly, in resolution 38/60 of 14 December 1983, had requested all States, the specialized agencies and other concerned United Nations organizations to co-operate actively in the Conference preparations. The Conference secretariat had recently dispatched a list of issues for comments by Governments. The issues listed included present status and future needs and priorities in the areas of nuclear power production and other peaceful applications of nuclear energy; existing or foreseeable constraints in the development of peaceful uses of nuclear energy; and practical measures and ways and means of promoting co-operation at the regional and international levels. The contributions to be provided by the International Atomic Energy Agency and other concerned organizations of the United Nations System, and the reports of the regional meetings, were to be submitted to the Preparatory Committee for the Conference for review at its next session to be held from 21 October to 1

November 1985 at Vienna. In this connection, the General Assembly, in its resolution adopted on 13 December 1984, had called upon the organizations to ensure that their input documents to the Conference, including the reports of the expert group meetings, should be concise and comprehensive and specifically related to the purpose, aims and objectives of the Conference, in accordance with General Assembly resolution 32/50.

The Secretary-General then pinpointed the major tasks of the regional preparatory work which included studies on regional experiences, problems and priorities in the peaceful uses of nuclear energy, followed by regional expert group meetings.

The Secretary-General stressed that the thrust of the Meeting would lie in identifying specific initiatives aimed at overcoming various identified constraints as well as practical measures and for promoting co-operation in the field

of peaceful uses of nuclear energy at the regional and international levels. To ensure effective co-ordination, appropriate consultative and review mechanisms might be required at both inter-organizational and intergovernmental levels; for instance, in the form of periodical meetings to review and monitor progress in the promotion of co-operation and to provide the necessary guidance and policy direction.

Ambassador Mehta concluded his address by reminding the Meeting of its special role as first in a series of regional expert group meetings, and of the region's role as a pace-setter in co-operative initiatives, such as the Regional Co-operative Agreement for Research, Development and Training Related to Nuclear Science and Technology (RCA) and a variety of other bilateral and multilateral arrangements in technical co-operation among the developing countries themselves.

# 8. PROCEEDINGS

The Expert Group took note of General Assembly resolution 32/50 of 8 December 1977 which declared that:

"(a) The use of nuclear energy for peaceful purposes is of great importance for the economic and social development of many countries;

(b) All States have the right, in accordance with the principle of sovereign equality, to develop their programme for the peaceful use of nuclear technology for economic and social development, in conformity with their priorities, interests and needs;

(c) All States, without discrimination, should have access to and should be free to acquire technology, equipment and materials for the peaceful use of nuclear energy;

(d) International co-operation in the field covered by the present resolution should be under agreed and appropriate international safeguards applied through the International Atomic Energy Agency on a non-discriminatory basis in order to prevent effectively proliferation of nuclear weapons".

By the same resolution, the General Assembly invited all States as well as the international organizations concerned to respect and observe the above principles as set forth in the resolution.

In the light of the above resolution, the Group examined the status and prospects for nuclear power and other peaceful applications of nuclear technology in the region; identified the major constraints; and suggested practical measures and effective ways and means for promoting international co-operation in the field of peaceful uses of nuclear energy for social and economic development.

The Meeting had before it document NR/ICPUNE/1 and Add. 1 and Add. 2 as a background information paper. In addition, working papers prepared by the experts representing member countries of the region were also available.

The Meeting commended the efforts made by the secretariat in providing a background document covering the application in the ESCAP region of nuclear energy for peaceful purposes in different socio-economic areas. In consideration of the additional information submitted by participants during the deliberations, the Meeting asked the secretariat to review and update the report taking into consideration also the information provided by various sources.

## A. Nuclear power in the ESCAP region

### A.1. Status of nuclear power

The countries of the ESCAP region showed a great diversity of physical size, population, resources and stages of development. Some were highly industrialized, some were partly industrialized and some had largely agricultural economies. Some had large and diverse physical, human and energy resources while others had little of these resources. A few of the countries were already benefiting greatly from the peaceful applications of nuclear power and nuclear research. Other countries were assessing the potential of nuclear energy to help meet their future energy and research needs. Some of the countries might not be interested in the use of nuclear energy for power generation in the foreseeable future because of their small size or the availability of alternative energy resources.

Five of the ESCAP countries had operating commercial nuclear power stations and four of these countries, with one additional country, also had plants under construction. Total installed nuclear power capacity in 1983 was about 25 gigawatts electricity (GWe), which represented 12 per cent of the world installed capacity, with a further 19 GWe under construction. Recent estimates by the International Atomic Energy Agency (IAEA) (Reference Data Series No. 1, 1984) indicate that there could be a total installed nuclear power capacity in the ESCAP region in 1990 of 41-47 GWe, which would represent about 11 per cent of world capacity. Several other ESCAP countries might consider nuclear power as an option in the 1990s. This could lead to a total installed capacity in the region of 80-125 GWe by 2000, which would represent about 15 per cent of estimated world capacity at that time.[1]

There had been considerable co-operation in the installation of commercial nuclear power and the use of research reactors in the region under a number of bilateral agreements and with the assistance provided by IAEA, but further developments in nuclear power generation in the region might be limited by a number of constraints which might also operate in other regions.

### A.2. Constraints to nuclear power development

The Expert Group felt that nuclear power had an important role to play in several countries of the region. In

---

[1] The figures of total regional capacities mentioned include those of Japan as follows: 19 GWe, 27-32 GWe and 50-76 GWe respectively in the years 1983, 1990 and 2000.

order to allow the realization of the potential of nuclear power in the region, it would be necessary to overcome a number of major constraints which were identified by the Expert Group as follows (constraints are not listed in any order of priority):

(a) Finance:

(i) High capital cost of nuclear power is an obstacle;

(ii) High import content has a negative impact on the balance of payments;

(iii) Mobilization of capital is thus a major problem.

(b) Infrastructure development and organization:

(i) The lack of assessment and planning capability makes it difficult to undertake the necessary long-term planning;

(ii) The lack of industrial support is a serious drawback in planning and implementation of nuclear power projects and causes problems during plant operation;

(iii) The lack of research and development support causes similar problems as above in regard to industrial support;

(iv) The inadequacy of safety and environmental regulations also causes problems during the implementation of projects as well as during plant operation.

(c) Manpower training:

(i) Shortage of qualified manpower for management, design, construction, operation and maintenance of power plants, and regulatory activities as well as for related research supports poses a serious problem.

(d) Transfer of technology:

(i) Transfer of technology is an issue under international discussion.

(e) Assurance of supply:

(i) Assurance of supply is also an issue under international discussion.

(f) Technical/economic:

(i) Small grid size and unavailability of suitable economic and small reactors prevent the use

of nuclear power in many national energy programmes;

(ii) Long construction time results in economic problems;

(iii) Problems connected with management of radioactive waste and spent fuel are constraints.

(g) Public acceptance:

(i) Lack of proper public awareness has a negative effect on decisions concerning introduction and utilization of nuclear power.

## A.3. Recommendations on nuclear power development

The Expert Group suggested the following practical measures and effective ways and means for the promotion of co-operation at the regional and international levels aimed at overcoming the above constraints:

(1) Keeping in view the important role that the World Bank and regional banks (such as the Asian Development Bank (ADB)) play in the development of energy resources in the developing countries, the Group recommended that these banks should include the nuclear power option in their country energy assessments and make provisions for long-term loans on easy terms for nuclear power plants in their energy financing programmes.

(2) International organizations should assist the countries of the region in setting up joint industrial ventures related to nuclear power generation to reduce foreign exchange outflows.

(3) International support should be provided for developing indigenous capability in assessment and planning, and in the design, construction, installation, operation and maintenance of nuclear power plants and related facilities by countries in the region.

(4) In the manpower development area, an essential part should be the setting up of new, and expansion of existing, international and regional training programmes, and the monitoring of their effectiveness. Topics for the training programmes should reflect the needs of the region.

(5) Taking note of the efforts being made in the areas of transfer of technology and assurance of supply, the Group expressed the view that no effort should be spared to find appropriate solutions to these issues on an international basis.

(6) The Expert Group recommended international initiatives to:

(a) Develop small- and medium-size power reactors at acceptable cost to developing countries with small grids;

(b) Explore the possibility of intercountry grid connections to increase grid size;

(c) Examine the feasibility of international and regional arrangements for nuclear fuel cycle and radioactive waste management facilities;

(d) Promote the standardization of nuclear power reactors to decrease construction time and capital cost;

(e) Improve distribution of information material on nuclear power to facilitate better public awareness.

## B. Nuclear techniques, radioisotopes and radiation

### B.1. Status of applications

#### a. *Agriculture* [2]

The ESCAP region embraced over half of the worlds population and the earth's land surface and included many agro-climates and aquatic conditions. The region offered suitable agro- and aqua-ecological conditions for the cultivation of practically all economic plant and animal species. Among the countries in the ESCAP region, seven fall into the least developed category, and six were highly developed in terms of indices of farm productivity and the use of superior agricultural technology. For the rest, low agricultural yields, and poor capital and technical inputs were common denominators. Increasing agricultural production and improving distribution, minimizing pre- and post-harvest losses, promoting better storage and longer shelf-life of food were important objectives.

Nuclear techniques could assist in an effective way in soil, water, fertilizer and pest management as well as agricultural and animal productivity, all of which were of major importance to the region. Since 1964 the Food and Agriculture Organization of the United Nations (FAO) and IAEA had jointly provided considerable assistance in applying nuclear techniques to developing food and agriculture in the ESCAP region. Since 1971, the assistance had been provided also through the Regional Co-operative Agreement.

#### b. *Hydrology*

Water management in the ESCAP region was a problem for agriculture, severely limiting crop yields. To increase agricultural production and minimize environmental degradation, it was essential that adequate water management methods be developed. Studies in isotope hydrology

---

[2] 'Agriculture' includes crops, forestry, livestock and fisheries.

were particularly important and had already been used extensively in some parts of the region, largely with the assistance of IAEA and several bilateral assistance programmes.

#### c. *Food and nutrition*

Food preservation was of vital importance in the ESCAP region and will help to overcome nutritional deficiencies. Between 15 and 30 per cent of food produced in the region was reportedly lost owing to spoilage, and therefore storage technology for food and extension of shelf-life had to be improved. Thirty years of research and development work on the preservation of food and disinfestation using ionizing radiation had shown that the treated food was safe for consumption by humans and animals, that the method was efficient, and that it required less energy than other preservation methods. The Expert Group took note of the significant and important developments internationally in food preservation as reported by the Joint FAO/IAEA/WHO Expert Committee on the Wholesomeness of Irradiated Food (1980) and the adoption of the Codex General Standard for Irradiated Foods by the FAO/WHO Codex Alimentarius Commission (CAC) in 1983.

#### d. *Medicine*

Recent developments had made radioisotopes and radiation essential tools in public health, medical diagnosis and therapy. While complex and expensive instruments had been introduced in some countries in the ESCAP region, there was still a great need for provision of more basic equipment and trained manpower to be able to deliver diagnostic and therapeutic services to the majority of the population, and in the maintenance of equipment. The technology had made significant contributions to public health, e.g. in the area of sterilization of medical supplies. The provision of radioisotopes was still in its infancy in many countries in the region.

#### e. *Industry*

While radioisotopes and radiation technology were being applied increasingly in industry in some developed countries, their use in the ESCAP region was still limited. They were capable of saving energy and resources, and their applications had contributed to increased industrial productivity and improved quality control.

### Summary

In summary, programmes for the use of nuclear science and technology in the areas of agriculture, food preservation, medical applications, hydrology and industry in countries in the ESCAP region varied from virtually non-existent to highly developed. A considerable potential existed to initiate new programmes and expand a number of existing programmes supported by United Nations agencies and by bilateral agreements.

## B.2. Constraints to applications

The Expert Group identified the following major constraints which had to be overcome to enable the full potential of nuclear science and technology to be realized (constraints are not listed in any order of priority):

(a) Finance:

(i) Inadequate finance for nuclear facilities, centres, demonstration projects, exhibitions, manpower training and information material.

(b) Infrastructure development and organization:

(i) Inadequate capability for maintenance of equipment and limited research and development facilities;

(ii) Inadequate international assistance to promote industrial applications in developing countries;

(iii) Inadequate international/regional mechanisms for scientific co-operation and exchange of material (e.g. germ-plasm and induced mutants).

(c) Manpower training:

(i) Inadequate international assistance for technical training for application of nuclear techniques;

(ii) Inadequate availability of training material for non-power applications of nuclear techniques.

(d) Transfer of technology:

(i) Inadequate technological support for complex equipment.

(e) Availability of supply:

(i) Inadequate supply of radioisotopes and sources of radiation.

(f) Public acceptance:

(i) Inadequate information material on the safety of irradiated food, and hydrological and industrial applications of nuclear techniques, and inadequate mechanisms for its distribution.

## B.3. Recommendations on applications of nuclear techniques, radioisotopes and radiation

The Expert Group suggested the following practical measures and effective ways and means for the promotion of co-operation at the regional and international levels aimed at overcoming the above constraints:

(1) Additional grants should be organized through the United Nations system. International and regional financial organizations, e.g. World Bank, Asian Development Bank, should provide loans on easy terms, to establish research facilities, demonstration projects, production plants, exhibitions, training and information material for applications of nuclear techniques in the region.

(2) ESCAP, in collaboration with other concerned international organizations (e.g. FAO, IAEA, World Health Organization, the United Nations Industrial Development Organization and the United Nations Development Programme) and relevant financial institutions, should take an active and effective role in mobilizing financial and technical resources, and in organizing and supervising such co-operative efforts as were considered necessary in support of member States in the region in the initiation and extension of nuclear applications in food and agriculture, animal sciences, industry and health towards social and economic development.

(3) International support should be provided to designated national and regional centres in member countries to undertake training programmes for the application of nuclear techniques in the region, and facilitate the transfer of technology in that field.

(4) International support should be provided to designated national and regional centres to co-operate in work on, and exchange of, material such as germ-plasm and induced mutants.

(5) International organizations should facilitate the setting up of national or regional capabilities for the production and distribution of a regular supply of radioisotopes from both reactors and equipment such as cyclotrons (a vital need was the provision of technetium-99m from its parent molybdenum-99 produced in reactors for medical applications).

(6) International support should be provided to assist national and regional efforts to provide adequate information material for the public, particularly on the safety of irradiated food and of hydrological and industrial applications using nuclear techniques. It is suggested that national food, health and legislative authorities should be assisted by international organizations in providing appropriate expert teams to help resolve issues of public acceptance.

### C. Concluding recommendation

The Expert Group recommended the establishment of an appropriate international mechanism to ensure proper and effective co-ordination of the co-operative activities as well as to review and monitor progress in the implementation of measures aimed at promoting international co-operation in the peaceful uses of nuclear energy for economic and social development.

### D. Adoption of the report

The Meeting adopted the report on 17 January 1985.

### Annex to this report
### LIST OF PARTICIPANTS

**Regional experts**

1. Manoon Aramrattana — Thailand
   Lee Gun Bae — Republic of Korea

G.B.A. Fernando — Sri Lanka

Mohd. Ghazali — Malaysia

Rosemary M. Greaves — Australia

Xu Guanren — China

C.J. Hardy — Australia

Arshad Muhammad Khan — Pakistan

Gary Makasiar — Philippines

Athorn Patumasootra — Thailand

Mohammed Abdul Quaiyum — Bangladesh

Budi Santoso Sudarsono — Indonesia

K. Sundaram — India

**Others**

Jean Jacques Graf — France

A. de Fleurieu — France

**Annex**

**REFERENCES, DATA SOURCES AND ABBREVIATIONS**

# REFERENCES AND DATA SOURCES

1. "Current and projected energy situation in the ESCAP region", E/ESCAP/NR. 8/18.

2. "Electric Power in Asia and the Pacific" (ESCAP, Bangkok).

3. "Nuclear power reactors in the world", Reference data series No. 2 (IAEA, April 1984).

4. "Energy in the Developing Countries" (World Bank, 1980).

5. Neels, Louis, "The realism of nuclear power", *World Science News*, vol. XIX, No. 14 (30 March – 6 April 1982), p. 12.

6. "Uranium", a Joint Report by OECD Nuclear Energy and IAEA (Organization for Economic Co-operation and Development, December 1983).

7. "Survey of energy resources", World Energy Conference, 1983.

8. *Mineral Yearbook 1975* (United States Bureau of Mines).

9. "Uranium and thorium : estimated production in 1978", *World Energy Conference Survey* (1979), table 7, part I.

10. FAO-RAPA Monograph No. 2, 1983 (FAO Regional Office for Asia and the Pacific, Bangkok).

11. Micke, A., *IAEA Bulletin,* vol. 26 (1984), p. 26-28.

12. Myint, U.T., *Proceedings of the Agricultural Research Congress* (Rangoon, Burma, 1981), p. 3-9.

13. Guang-chang, S., *New frontiers in technology application* (Tycooly International Publishing Ltd., Dublin), 1983, p. 151-156.

14. *Production Yearbook 1981* (FAO, Rome, 1982).

15. "Maximizing the efficiency of fertilizer use by grain crops", *Fertilizer Bulletin,* No. 3 (FAO, Rome, 1980).

16. Agricultural Division, ESCAP.

17. Mukerjee, IAEA-SM-263/35, 1982.

18. Banerji, A.K., Pesticides, 3, 1981.

19. FAO Agricultural Service Bulletin, No. 43.

20. "Wholesomeness of irradiated food", Report of a Joint FAO/IAEA/WHO Expert Committee, WHO Technical Report Series, No. 659 (1981).

21. "Recommended international general standard for irradiated foods and recommended international codes of practice for the operation of irradiation facilities for the treatment of foods", Joint FAO/ WHO Food Standards Programme, Codex Alimentarius Commission, CAC/RS 106-1979, CAC/RCP 19-1979.

22. van Kooij, J., *IAEA Bulletin,* vol. 23, No. 3 (1981).

23. *IAEA Bulletin,* vol. 23, No. 4 (1981).

24. IAEA Technical Assistance Report 1983.

25. Kobayashi, M., *IAEA Bulletin,* vol. 26, No. 1 (1984), p. 22-27.

**ABBREVIATIONS** (with the exception of operators and contractors of nuclear power reactors in Part 1., table 1.4).

| | |
|---|---|
| ADB | Asian Development Bank |
| Agr, Agrl | Agriculture, agricultural |
| BWR | Boiling light-water-cooled and moderated reactor |
| °C | degree Celsius |
| CAC | Codex Alimentarius Commission |
| Ci | Curie(s) |
| C-14 | Carbon-14 |
| Co-60 | Cobalt-60 |
| Cs-137 | Caesium-137 |
| EDB | Ethylene dibromide |
| FAO | Food and Agriculture Organization |
| FBR | Fast breeder reactor |
| GCR | Gas-cooled reactor |
| GW | Gigawatt (Giga, G, billion, $10^9$) |
| GWe | Gigawatt electricity |
| ha | hectare |

| | |
|---|---|
| HWLWR | Heavy-water moderated, boiling light-water-cooled reactor |
| H-3 | Tritium |
| IAEA Agency | International Atomic Energy Agency |
| IARI | Indian Agricultural Research Institute |
| INFCE | International Nuclear Fuel Cycle Evaluation |
| I-125, I-131 : | Iodine-125, Iodine-131 |
| k, kg | kilo (1 000), kilogram |
| kGy | kiloGray (10 kGy = 1 Megarad, 1 Mrad = $10^6$ rad) |
| kwh | kilowatt-hour |
| MeV | Million electron volts |
| MW | Megawatt (mega, M, million) |
| NDT | Non-destructive testing |
| NEA | Nuclear Energy Agency (of OECD) |
| N-15 | Nitrogen-15 |
| OECD | Organization of Economic Co-operation and Development |
| PHWR | Pressurized heavy-water-moderated and cooled reactor |
| ppm | part per million |
| PWR | Pressurized light-water moderated and cooled reactor |
| P-32 | Phosphorus-32 |
| Pb-210 | Lead-210 |
| RCA | Regional Co-operative Agreement for Research, Development and Training related to Nuclear Science and Technology (of IAEA) |
| Rs | Indian Rupees (in this document) |
| SIT | Sterile insect technique |
| SSDLs | Secondary Standard Dosimetry Laboratories |
| S-35 | Sulphur-35 (Sulfer-35) |
| Sn-113 | Tin-113 |
| Th | Thorium |
| T-3 | Tri-iodo-L-thyronine |
| T-4 | Thyroxine |
| UNDP | United Nations Development Programme |
| UNIDO | United Nations Industrial Development Organization |
| $U_3O_8$ | Chemical symbol for a mixture of uranium oxides |
| WEC | World Energy Conference |
| WHO | World Health Organization |